ENVIRONMENTAL POLLUTION

Yogendra N. Srivastava

A P H PUBLISHING CORPORATION
4435-36/7, ANSARI ROAD, DARYA GANJ
NEW DELHI-110 002

Published by
S.B. Nangia
A P H Publishing Corporation
4435-36/7, Ansari Road, Daryaganj
New Delhi 110002
Ph.: 23274050
E-mail : aphbooks@gmail.com

2026

Printed at
Balaji Offset
Navin Shahdara, Delhi 110032

PREFACE

Concern for environment is a matter of one and a half decades only. We all wish to live in an environment free from any kind of contamination. But very few of us are aware and concerned about the environment. This text is based on larger issues of environment at national level. This book provides fundamental concept and other related matters regarding environmental pollution in a nutshell.

The basic object of this book is to develop creative awareness and concern for the environment among the students as well as general reader.

I am grateful to my wife and children who extended me full co-operation in completing this endeavour.

YOGENDRA N. SRIVASTAVA

Gorakhpur

CONTENTS

Chapter | *Page*

1

INTRODUCTION

"The more we exploit nature, the more our options are reduced, until we have only one to fight for survival."

—MORRIS K. UDALL

"We won't have a society if we destory the environment."

—MARGRET MEAD

"Every second hectare of the India's 304 hectares faced some kind of ecological problem, a large percentage of drinking water provided by India's major rivers is polluted and clean air is breathed by Indians for only two hours in the morning. Over 10 per cent of hospital patients are victims of a polluted environment."

The above facts were brought to light during the debate on the 'Impact of Development of Science and Technology on Environment' at the 68th session of the Indian Science Congress in Varanasi.

The effect of Environmental pollution is not only on India but encompasses the whole world. According to a survey conducted by United Nations the concentration of carbon dioxide in the atmosphere is now 330 PPM, which is 14 per cent more than the concentration of carbon dioxide ten years ago. In India the problem of environment and pollution became an important issue only very recently. But if we trace history we find that concern for the environment has been part of India's social and cultural heritage. This land of ours is the birth place of sages and of religions, preaching love for every living being, a concern for life and conservation of nature. Twenty two centuries ago, Emperor Asoka emphasized to preserve animal life and forest trees.

The concern for environment in present time found expression in proclamation undertaken to "Defend and improve the environment for present and future generation" by the United Nations Conference on Human environment in Stockholm in 1972.

All natural eco-systems are balanced and were unaffected by man until the dawn of civilization. But the speed and nature of man-induced environmental changes have brought an increasing disharmony between him and the nature at present. Living beings are dependent upon nature for their requirements. Hence the purity and balance of environment is essential. Some of the problems such as depletion of resources, soil erosion, pollution, deforestation and extinction of wild animals due to increased human population have reached global level.

Man has played a very important part in shaping his environment. He has been responsible for degrading the quality of his environment ever since he appeared on this earth. At first he contaminated the atmosphere by the use of fire which added gases, smoke and ash to it. When he came out of the cave age and began to settle into villages, towns and cities, the situation gradually worsened. The degradation of the environment was caused mostly by his activities such as burning of wood, smelting of ores, tanning leather, primitive methods of sewage disposal and so on. With the advent of industrial revolution, coupled with urbanization, all kinds of impurities began to be added to the natural air, water, as well as soil, causing almost irrepairable damage to environment.

diverse range of pollutants such as gases, particulates, agricultural chemicals and radioactive materials in the atmosphere;

oil spills, soil wastes on land are affecting the organisms directly or indirectly. Infact, the pollution has assumed distressing dimensions for the present as well as the coming generations. The government of almost all the countries of the world are forced to pay serious attention to not only conserve and improve the environment but also to prevent it from further deterioration.

ENVIRONMENT

Environment is the sum of substances and forces external to the organism in such a way that it affects the organism's existence. In relation to man, the environment constitutes of air, land, water, flora and fauna because these regulate the man's life.

Environment is a multi-dimensional systems of complex inter-relationships in a continuing state of change.

By environment we mean not only our immediate surrounding but also a variety of issue connected with human activity, productivity, basic living and its impact on natural resources such as land, water, atmosphere, forests, dams, habitat, health, energy resources, wild life, etc.

Like other animals man depends on environment and becomes an environmental factor with respect to other members in an eco-system.

POLLUTION

The word 'Pollution' is derived from a Latin Word 'Pulluere' which means 'to soil' or 'to defile'.

Pollution means an undesirable change in physical, chemical or biological characteristics of air, land and water that may or will harmfully affect the human, animal and plant life.

Contamination of environment with impurities making it unfit for its intened use is known as pollution.

Pollution may be defined as contamination of air, water or soil with undesirable amounts of material or heat. Heat is not a pollutant but change in heat means change in climatic condition.

Pollutant—A substance the presence of which causes pollution is known as pollutant.

Pollution is deterioration of the quality of environment by the production of the quality of impurities.

There are four main kinds of pollution as under :

(*a*) Air Pollution,

(*b*) Land Pollution,

(*c*) Water Pollution,

(*d*) Noise Pollution.

Radiation has caused hazards in recent times this may be termed as Radiation Pollution'.

Air is a gaseous mixture of oxygen 21 per cent, Nitrogen 78 per cent carbondioxide and Inert gases 1 per cent.

Oxygen present in the air is essential for life and is known as the life supporting gas. When we breathed in air, air enters our lungs through the nose and windpipe. This is known as inspiration. In the lungs exchange of gases takes place where oxygen is passed into the blood for oxidation of food and the waste gases are breathed out in the reverse direction.

Nitrogen present in the air is also useful to us. When it is captured and converted into nitrogenous substances like nitrates in soil, the soil is enriched. This is essential for the healthy growth of plants.

Carbondioxide present in the air is essential for plant life because plant absorb carbondioxide from air and release oxygen into the air, which is essential for animal and human beings. Hence we can say that plants and animals are dependant on each other and owe their existence to air which is a great gift of nature. The proportion of oxygen, nitrogen and other gases in air is fixed and definite. Any change in the proportion would affect the life of plants, man and animals.

This change is brought about by man. Industries, vehicular traffic, domestic usage of fuel etc., release hazardous emissions when these emissions are absorded into the atmosphere the composition of air is changed.

Air Pollution

History of air pollution could be traced back to the industrial revolution and the discovery of steam engine, in the eighteenth century. Increased use of coal and the smoke and sulphur compounds emanating from it began to contaminate the atmosphere more and more.

The first major air pollution study in the United States and very important study of chicago was in 1912-1915, largely at the instigation of one Thomas Donn'ielly. He had a printing plant immediately adjacent to one of the main rail road terminals and found that on many days, his freshly printed works were rendered unsaleable by the layer of root that landed them. He tried to begin a movement to electrify the railroads before they entered chicago, as New York had already done. Sensing a threat, the local industries turned this into a major study of air quality in chicago and all other aspects of problem. In the short run, in their study, the particulate coating of atmosphere was collected on filters and weighed.

A serious problem has arisen now, especially in the congested and industrialised cities, with heavy vehicular traffic. Most of the pollutants are in gaseous form, and these are explained in Chapter 2.

DEFINITION OF AIR POLLUTION

The World Health Organisation has defined air pollution as "substances put into air by the activity of mankind in concentration sufficient to cause harmful effect to his health, vegetables, property or to interfere with the enjoyment of his property."

DEFINITION OF AIR POLLUTION GIVEN BY THE STATE OF NEW JERSEY

"Air pollution is the presence in the outdoor atmosphere of one or more air contaminants in such quantities and duration as are or tend to injuries to human health or, welfare, animal or plant life or would increasingly interfere with the enjoyment of life and property."

Engineer's Joint Council for Air Pollution and its control, USA, defines Air Pollution as :

"Air pollution means the presence in the outdoor atmosphere of one or more contaminants, such as dust, fumes, gas, mist, odour, smoke, or vapour in quantities, of characteristics, and of duration, such as to be injurious to human, plant or animal life or to property, or which unreasonably interferes with the comfortable enjoyment of life and property."

Once in the atmosphere, these pollutants cannot escape. The roads, bricks and stones absorb and hold the heat during daytime creating heat islands during the nights. The pollutant containing air keeps on circulating over the city. Since there is an intimate relationship between solar radiation, air, moisture, soil and plants the polluted environment greatly, affects the growth and development of these primary producers.

The following are the meteorological variables which affect severity of a pollution :

(*a*) Wind speed and direction;

(*b*) Atmosphere diffusion;

(*c*) Temperature variation with height, includes lapse rate and Inversion;

(*d*) Mean maximum Depths;

(*e*) Precipitation.

The following are the non-meteorological factors which also affect the air pollution :

(*a*) Topographical features.

(*b*) Quality and Quantity of pollution.

(a) Wind Speed and Direction

Pollutions are dissipitated in the atmosphere by both the horizontal and vertical movements of the wind. In general, greater the wind velocity, the greater is the dilution. Turbulence is a stirring action of wind and, therefore, is related to the vertical movement of wind moving across the surface of earth is fluid moving along the surface of the containing structure. The drag against the surface of the earth causes the wind near the surface to blow at very slow speed. Wind speed increases very rapidly with height upto 10 metre. Above this, it continues to increase in speed with height but at a much lower rate.

(b) Atmospheric Diffusion

Atmospheric diffusion is the movement of large parcels of air from one point to another.

(c) Temperature Variation

The temperature distribution of the atmosphere depends upon the rate at which energy being received from the sun and on various transport mechanism *i.e.*, electromagnetic radiation, convection, evaporation, etc. As a consequence the temperature is not constant but varies with height, season, time of day, amount of cloud cover and many other variables.

(*i*) *Lapse Rate*

Lapse rate is defined as equal to the adiabatic lapse rate minus the temperature lapse rate. Temperature Lapse Rate is the ratio of change in temperature to change in height in 100s of metres. The adiabatic lapse rate is the rate of cooling with lofting (or heating upon discent) of a parcel of air with no heat exchanges. A parcel of dry air expands upon rising. When air expands, it cools at the adiabatic lapse rate if no heat exchange occurs. If the surrounding air is cooler, the parcel will continue to rise and therefore, unstable conditions or strong lapse rate exists. A strong lapse rate indicates instable conditions and therefore, good air mixing. More positive lapse rate indicates stable air, which means air pollution will not be removed because there is not sufficient atmosphere turbulence or mixing. Inversion of the temperature lapse rate *i.e.*, from positive to negative, resulting in stable atmospheric conditions is known as 'Inversion'.

(*ii*) *Temperature Inversion*

Temperature inversion may be said to be a meteorological condition in which air pollutants are unable to rise and be disbursed in the atmosphere and produce high concentration of pollution.

A temperature inversion is said to exist when the normal lapse rate in the lower atmosphere is invested and the temperature actually increases with height, *i.e.*, exhaust gases and pollutants. Which rise only a certain distance under these conditions and persistent inversion that last several days can be dangerous.

Inversion occurs frequently during the night or early morning hours. Low wind speeds, equal or less than 10 kmph usually accompany inversions, so there is very little horizontal dispersion of pollutants. Inversion temperatures are usually limited to the first 500 metres and this, therefore, is the maximum inversion heights.

(d) Mean Maximum Depths

The mean maximum depth (MMD) is the height to which the unstable air mixes. In the absence of radio sound observations, the maximum mixing depth may be estimated as being the height of the loop of the low altitude cloud layer. The maximum mixing depth varies during the day as well as varies from season to season. Variation is also dependent upon the topographical features. Vertical depression of pollutants is limited by the ground and the MMD. Therefore, mixing depths are essential in estimating the amount of vertical diffusion of pollution in the atmosphere.

(e) Precipitation

Pollutants can be washed out of the air by the natural secrubbing action of rain, snow and all other forms of precipitations as it falls to the ground. Gases that are soluble in water are removed by adoption when the particles stick to the precipitation after being impacted by it. Not only can precipitation affect the atmospheric pollution, but the pollution can affect the precipitation.

(f) Topographical Features

Topography can very seriously affect local atmospheric conditions. Most important of these are valleys, shore-lines and hills. Most of the infamous air pollution episodes occurred in locations with adverse topographical conditions. Valley affects occur due to the channelling of winds. Valleys tend to make the windflow in the general direction of the valley axis. Slope winds occur in valleys in the evening when the air near the ground is cooled. Any air pollution relased on the slopes or in the bottom of valleys will stay in the valley if an inversion also exists.

Shore-line winds are created because of the differences in heating rates of the earth and the water under the same amount of the sun shine land absorbs heat faster than water. Therefore, on a sunny day, sea breeze are created which come from water to the land. Land also cools faster than water creating on cloudy days or at night times land breezes. Land breezes are usually lower in velocity and shallower than sea breeze.

Hills can cause a varying degree of influence on air pollution removal. A smooth hill causes the least affect of the flow of air. Rough hills conversely cause turbulence eddies and good mixing which promote improved air pollution removal.

(g) Quality and Quantity

Quality and Quantity of air pollution can be estimated with regard to different sources of pollutions such as transportation, industry, power generation, space heating and refuse burning. Approximately 90 per cent by weight of this pollution is gaseous and 10 per cent particulate matter. The pollutants from the different sources can be further classified according to their physical and chemical composition such as inorganic or organic gases and particulates. However, there are no correct measurements that can tell about the emission of pollutants from the same source under different processes. The ability to determine the effectiveness of controlling a number of pollutants is likely to depend on our measurements.

The air pollution problems of one city or country may differ greatly from those of another city or country. This is true even if the sources of air pollution are similar. Meteorology plays a very important role in determining the significant level of air pollution.

2

AIR POLLUTANTS

Contaminations in the air are present either in gaseous form or as particles.

GASEOUS POLLUTANTS

1. Sulphur Oxides, Hydrogen Sulphides

These are related by the biological decomposition and from volcanic eruptions. Smelting of sulphide-containing ores, combustion of sulphur-containing fuels such as coal and oil, petroleum refining and geothermal energy sources are some of the important contributors. While high concentrations of sulphur dioxide cause intevenial and blade damage, necrosis *i.e.*, death of living tissues of leaves and cellular collpase, lower amount result chlorosis *i.e.*, loss of green colour of the leaves without cellular collapse, still lower quantity suppresses the overall vegetative as well as reproductive growth and yield. Exposure to hydrogen sulphide results in leaf lesions, defoliation and reduced growth. Sulphur dioxide paralyses or destroys bronchial cilia in air passage of man, constricts bronchiae, damages lungs, lowers resistance to pneumonia and influenza and causes bronchitis, emphysema and irritation of mucous membrance.

2. Hydrogen Fluorides

Active volcanoes are the natural sources of fluorides in the atmosphere. These are also emitted from aluminium, steel and electrochemical reducation plants, blast furnaces, brick, tile and superphosphate fertilizer industries and from the combustion of coal.

Fluroride burns the tip of leaves. Low amounts impair plant growth, result in excessive dropping of bloom and fruits, development of small, partially or completely seedless fruits, and premature formation of soft red flesh and spliting of peach

3. Nitrogen Oxides

Anaerobic breakdown of nitrogenous compounds by bacteria, forest fires and lightening constitute the natural causes. Chief sources, however, are power generators and motor vehicles. Burning of organic wastes and manufacture of explosives and nitrogenous fertilizers further add to this problem.

Nitrogen dioixde brings about bifacial necrosis leading to collapse of leaves, enhancement of green colour followed by chlorisis and extensive leaf drop, as well as increase in fruit yield In men, it produces general and pulmonary oedema and haemorrhage.

4. Hydrogen Chloride

It is infrequently expelled by accidental spills from chemical manufacturing plants. Besides, it is released from combustion of coal, paper, plastics and chlorinated hydrocarbons, and ignition of solid-fuel rocked engines.

Hydrogen chloride is responsible for the abaxial glazing of leaves caused by collapse and plasmolysis of epidermal cells. High concentrations produce nectrotic lesions.

5. Hydrocarbons

Biological decomposition of organic matter, seepage from natural gas and oil-fields and volatile emissions from plants are major causes for the release of hydrocarbons such as methane, ethylene and aniline. Incomplete combustion of fuels, motor vehicles exhaust, petroleum-refineries, agricultural burning, motor fuel marketing, manufacture of explosives and cracking of natural gas in petrochemical plants constitute, the anthropogenic sources that emit hydrocarbons.

Ethylene causes yellowing and occasional necrosis of leaves, chlorisis of floral buds, inhibition of terminal growth, shortening of internodes, thickness of stems, lack of apical dominance, reduced growth, dry sepal diseases of orchids and decrease in amount of chlorophyll and cardenoids. In man, hydrocarbons bring about irritation of mucous membrance, bronchial constriction and eye irritation some of these have been shown to trigger development of lung cancer in experimental animals. We are also aware of the harm done by methyl Isocynate which accidently leaked out from the storage tanks of the pesticide factory in Bhopal on December 2, 1984 killing over 3 000 people and seriously affecting lakhs of residents.

6. Ammonia

Refrigerator pre-cooler system of cold storage, manufacture of anhydrous amonium fertilizers and nitric acids, domestic incinerations are the prime generators of amonia.

It induces bleaching of leaves rusty spots on leaves and flowers, reduction of root and shoot growth, browning and softening of fruits, development of dark, corky lenticels in apples and reduction in the rate of seed germination.

7. Carbon Monoxide

It is released chiefly from automobile engines, and defective furnaces. Being colourless and odourless, the presence of carbonmonoxide gas cannot be sensed and we unconsciouly inhale it with the polluted air that we breathe, carbonmonoxide damages body cells and poses serious health hazards. It provides headache, dizziness, inability to distinguish time intervals and other observable physical effects, and cardiac and pulmoneary changes. It forms carboxy—haemoglobin in RBC which prevents them from carrying oxygen to all parts of the body. As per Indian standard Institute emission by motor vehicles should not exceed three per cent and from five year old vehicles it should not be more than 4.5 per cent.

8. Photochemical Oxidants

These are secondary pollutants which are mainly formed by photochemical reactions between primary pollutants-nitrogen oxides and hydrocarbons. The major oxidants are ozone and peroxyacetylene nitrate (PAN). Minor amounts of ozone are also added to the atmosphere by electrical discharge such as lightning flashes, by vertical flux of stratospheric ozone and by tropospheric electric storms. Ozone is a natural constitutent of upper atmosphere *i.e.*, stratosphere, where it is formed and destroyed in a cyclical process with sunlight as the driving force. Most of the solar ultraviolet radiation is absorbed by stratospheric ozone before it reaches the lower atmosphere *i.e.*, troposphere, where approximately ten per cent of the atmospheric concentration of ozone is found. Ozone is also produced in rest quantities in the atmosphere of every big city by the action of sunlight on various waste products of combination. Ozone concentrations are increasing at an alarming rate, threatening man's health and the productivity of his crops. It has been calculated that doubling the troposheric ozone content may raise the surface temperatures by 1°C, which is substantial compared to the 2°C—3°C temperature rise calculated for a doubling of carbondioxide. It has also calculated that man's activities may lead to the doubling of the troposheric ozone content by the end of the century. Ozone filters the ultraviolet rays coming from cosmicrays to earth.

Minor amounts of ozone are added to the atmosphere by electrical discharges such as lighting flashes. More may be brought down from upper atmosphere by vertical flux, But the truly significant quantities of ozone present in our immediate environment are formed chemically by the action of ultra violet light on Nitrogen oxide. A multitude of combustion processes, particularly the inefficient internal combustion engines of automobiles daily emit tonnes of waste hydrocarbons and Nitrogen oxides into the atmosphere. Heat from any flame causes atmospheric nitrogen and oxygen to combine into nitrogen oxides. The hotter the flame, the greater the production of nitric oxides. The Nitric oxides is oxidized to nitrogendioxide by atmospheric oxygen but energy from sunlight quickly splits the nitrozen dioxide back into nitric oxide and atomic oxygen, which combines with the molecular oxygen of the atmosphere to form ozone.

Ozone generates necrotic flecking of upper surface of leaves, general chlorisis and bronzing, precious dropping of olderleaves, reduced growth of shoots and roots, suppression of nodulation, reduction in seed set and depression of marketable yields. It brings about shrinking of nuclei and cytoplasm of mesophyll cells which becomes granular, causing an increase in inter cellular space.

Ozone is now known to be cause of several widerspread diseases such as weather fleck of tobacco, leaf tip burn of carnations, tip burns of onions, bronzing of beam, speckle leaf of potato, and brown leaf of grapes. PAN results in bronzing and glazing of abaxial leaf surface which is due to plasmolysis and collapse of mesophyll cells around substometal chambers. Epidermal and guard cells are not injured. In human beings, oxidants cause stinging of eyes, coughing, headache, severe tired feeling, pulmonary congestion, oedema, haemorrhage, dry throat, disorientation altered breathing patterns. narrowing of airways and ageing of lung tissues.

9. Tobacco Smoke

It is mainly produced by smoking cigarettes and bidis. It is gradually becoming a potent pollutant especially in closed atmospheres such as buses, trains, auditoria and so on. It is also suspected of causing lung cancer, pulmonary and coronary heart diseases. It brings about thickening of bronchial epithelial layer, loss of ciliated cells, and appearance of cells with bizzare nuclei, which are probably the percursors of cancerous cells.

PARTICULATE POLLUTANTS

These pollutants are found in environment as particles ;

1. Fluorides

The particulate fluorides originate in the same way as the gaseous fluorides. However, these are less phytotoxic harmful or toxic to plants. They cause an increase in the fluoride content of the leaves and occasional tip burn. But the growth and yield are not greatly affected. The ingestion by cattle of various fluorine compounds falling on for ages, cause abnormal calcification of bones and teeth called flurosis, eventually resulting in loss of their weight, etc.

2. Lead

Lead is one of the toxic trace elements present in the human body. A trace element is termed essential if its deficiency results in the impairment of a function. Lead comes in the category of those trace elements which are known to be toxic even when present in traces and reportedly enter our bodies as a result of environmental pollution. Examples are lead, cadmium, mercury etc. Out of the known trace elements, the more toxic ones fall under the group called, heavy metals of which lead is one.

SOURCES OF ENVIRONMENTAL LEAD

Lead belongs to the group of those heavy metals which are extensively used by man. It is one of the softest and heaviest metals available. Its chief source is tetraethyl lead used as an antiknock additive to gasoline to increase its 'octane number'. The lead-containing gasoline fumes from automobile exhausts constitute the chief and widespread source of lead contamination in urban environment. That is why its concentration is higher in urban areas where automobile and industrial exhausts are more. Lead is widely used as a material of choice for shielding against x-rays and other nuclear radiations. Lead arsenate finds use in insecticides and lead borate in plastics industry. Significant deposition of lead also comes from smelting complexes, ceramics, paints, pesticides and solder used for scaling. Lead accumulates in considerable amounts in the leaf tissues as also in the tissue of human body.

EFFECT

Lead interferes with development and maturation of Red Blood corpuscles in human body. Even 0.2 parts per million (PPM) concentration creates metabolic disturbances, chronic exposure leads to stripped red cells. Being a cumulative poison it disrupts the functioning of cells and organs of the muscular, circulatory and nervous system for binding with the cellular enzymes also causing coagulation of proteins. It damages liver, kidney and gastro intestine and induces abnormalities in fertility, and pregnancy.

3. Cement Kiln and Other Dust

Particulates emitted from cement manufacturing units bring about premature fall of needless, higher pubescene of leaves, formation of more stomata and trichomes reduction in number and sizes of cobs and weight of seeds and increase in number of infertile seeds. Dusts from stone crushers, lime kilns, slate making units and quarries are not less hazardous.

4. Potassium Salts

Derived mainly from potash mines, they cause branch tip death, chlorosis and necrosis of leaves.

5. Sodium Chloride

The de-icing salts, mainly sodium chloride, used to remove ice and snow in winters, have been recognised to cause damage to the roadside trees in the form of leaf necrosis, defolitation, supression of flowing, and die back of terminal shoots in apple.

6. Agricultural Chemicals

Several chemicals such as insecticides, herbicides, fungicides and pesticides used widely in agriculture, are known to produce foliar lessions, chlorosis and abscission of leaves and reduction in fruit set. Several other types of particles in air such as coal dust and asbestos have been linked with necrotic lessions, reduction in fruit set in plants, silicosis and lung cancer respectively.

Living organism are rarely, if ever, exposed to single pollutant in nature. The influence of mixtures of pollutants is usually synergistic *i.e.*, greater than the additive effect of the different pollutants alone.

INFLUENCE OF AIR POLLUTION

The major factor influencing the climate of the world is the earth's heat balance. The amount of heat received, retained and reflected by the earth and its atmosphere determines the state of this delicate ecological balance upon which the health and well being of people on earth depend.

Our atmosphere contains a definite proportion of different gases. A clear environment constitutes 0.78 per cent Nitrogen, 0.03 per cent carbon dioxide and 16 per cent oxygen by volume. Besides it also contains ozone, Hydrogen Sulphide, Sulphur dioxide and carbon monoxide gases. If the ratio of any of these gases is disturbed the environment becomes polluted.

GREEN HOUSE EFFECT

In 1961 John Tindal a scientist, brought to the knowledge the three Green House Effects of carbon dioxide they are as follows :

(*a*) The excess concentration of carbon dioxide increases the heat of atmosphere.

(*b*) Carbon dioxide is transparent for solar radiation *i.e.*, it helps ultraviolet and Infra red radiations to reach on the earth.

(*c*) Carbon dioxide is opaque for the thermal radiation and therefore, absorbs the heat of the earth. Hence the temperature of earth is increasing.

The above effect of carbon dioxide is termed as 'Green house effect'.

Although carbondioxide is usually not considered a pollutant its concentration in the atmosphere has been increasing in recent years, due to enhanced combustion fossil fuels. Since this gas strongly absorbs the heat energy, it retards the radiative cooling of the earth. For example, its concentration increased from 290 parts per million in 1870 to 322 parts per million in 1970. Theoretically this should result in an increase of temperature by 0.5°C towards the year 2000. This warning portends melting of glaciers and flooding of the populous coastal plains. Since at present, earth appears to be rather cooling and not warming, it is impossible to predict the total effects of such a temperature change on the world climate.

There have been a lot of complaints in Britain about the wet summer weather experienced this year. But according to a team of scientists from England and the USA, the climate is likely to become wetter still. It is due to the so called "global greenhouse effect" which changing the pattern of rainfall. Temperature areas like Europe are getting wetter, while the long-term forecast for Africa and the American mid west is drought.

The "greenhouse effect" is caused by increasing levels of carbondioxide in the atmosphere, largely the result of burning wood and fossil fuels like oil and coal. The carbondioxide acts as a giant greenhouse around the globe. The sun's infra-red rays warm up the earth and the heat is then trapped by the carbondioxide and prevented from escaping into space. As the carbondioxide levels increase, this process is intensified.

Since the industrial revolution, the amount of carbondioxide in the atmosphere has increased by about 25 per cent. As a result the earth's temperature has gone up by about half a degree celsius in the past 130 years.

The biggest changes have taken place in the past 40 years. Areas in latitudes between 35 and 70 degrees north—that is much of Europe, USSR, Canada and part of China—are getting wetter. Areas in sub-tropics (5 to 35 degrees north) are getting drier. The precipitation aroused the equator is changing very little.

Europe has been getting steadily wetter since the 19th century with an even more dramatic increase in the Soviet Union. Rainfall in the United States declined from about 1880 to 1940, but has been on the increase since. Most significant of all levels of rainfall have declined drastically. It is interesting to know that over the past 100 years the greatest rise in the average temperature in the Northern hemisphere occurred before 1940. In the South the South the greatest temperature rise has occurred in recent years.

Similarly, in recent years, ozone, that screens out more than 99 per cent of solar radiation at wavelengths shorter than 320 nanometers has been threatened by a wide range of human activities. The earth is the only planet in the universe which has life on it. One of the various reasons ascribed to birth of life on earth is due to the protection from the deadly ultraviolet radiations of the Sun. Earth was able to make ultraviolet rays ineffective by its absorption by ozone.

Ozone a gas composed of three oxygen atoms, surrounds the Earth in the form of a thin cell, and is found upto the heights of 60 km. about the Earth's surface. It is most dense at 20-25 km. height, but even here only one molecule in 1,00,000 is ozone. Ozone is produced naturally, from oxygen, high in the atmosphere. Natural forces also break it down, the result that the gas is constantly being created and destroyed. Several chemicals used in or produced by industry greatly affect the speed at which ozone is broken down. It seems that chloro-flurocarbons which are used as propellants in aerosols released from spray cans in refrigeration technology, and other applications, rise into the stratosphere 15 to 50 km. above the earth. Here they are broken down by sunlight and give up chlorine which can destroy stratospheric ozone significantly. Such an ozone depletion on a global basis is likely to cause increase in skin cancers, suppress the immune system of humans, reduce crop yields and alter acquatic eco-systems which affects the benefits derived from marine sources. Ozone depletion also causes warming of seas. In view of the predicted adverse effects on the environment due to ozone layer depletion on a global scale, the forty countries which met at Montreal recently agreed to restrict the use of chloroflurocarbon compounds and related chemicals that deplete the ozone layer in stages as protective measure to curb the atmospheric levels of chlorine and Bromine. A worldwide fleet of supersonic transport SST planes increased use of aerosols, and refrigerants, and large scale of nuclear explosions cause a depletion of the stratospheric ozone. This enhances the level of ultraviolet radiation reaching the earth's surface. High intensities of the ultraviolet radiation are harmful to nearly all forms of life. The most alarming effect of the increase in ultraviolet radiations might be an increase of skin cancer. The influence on vegetation is less clear. It causes stunted growth, short, thick stems, smaller leaves, plasmolysis of cells, destruction of authocyarin, chlorophyll and nuclei, Bronzing of leaves, injury to various fruits, are some other implications. It also results in somatic aberrations, discolouring of staminal hairs, inhibition of pollen germination and pollen tube growth.

ACID RAINS

Acid rains means in common language the presence of excessive acids in rain water. It is one of the effects of air pollution. Every source of energy that we use—be it coal, fuelwood or petroleum products, contain sulphur and nitrogen. These two elements when burnt in the presence of atmospheric oxygen get converted into their respective oxides-sulphur dioxide and nitrogen dioxide, which are highly soluble in water. During rain these oxides react with large quantities of water vapour of the atmosphere and produce acid like sulphuric acid. Sulphurous acid, nitric acid and nitrous acid which then return to the earth's surface, with rain water.

The acids are very dangerous to the living organisms as they can destroy life. Acid rain can play havoc with the human nervous system by making the person an easy prey to neurological diseases. This because these acids produce highly toxic compounds which contaminates the potable water and enter our body.

Acid rain is already an acute problem in North America and Europe. Crops and forests in Canada are being destroyed by acid rain due to pollutants emitted by industries in Northern USA. Acidity Kills fish, bacteria and algae and the acquatic eco-system is destroyed.

Winds carry air pollutants from one country to another. Air pollution in England now descends upon sweden as acid rain.

Acidification of soil changes its biology and chemistry. When the soil becomes acidified, plants can absorb cadmium more easily and high levels of cadmium in plants is dangerous for animals and human beings.

Acid rain in Japan has damaged 5,000 sq. kms of cedar trees in Kanto plain which lies to the north of Tokyo. This area has high acid deposition brought about by air pollutants.

Acidic air pollutants are responsible for many other damaging effects such as corrosion of metals, weakening or disintegration of textiles, paper and marble. Investigations are going on to find out whether Taj Mahal is being affected due to pollutants released from Mathura refinery. Hydrogen Sulphide tarnishes silver and blackens leaded house paints, ozone produces cracks in rubber; its economic significance is apparent.

Air pollution can cause or contribute to a variety of safety hazards, *e.g.*, hazards associated with reduced visibility owing to smog etc. Air contaminants that are detectable by the sense of sight, touch, smell or taste can be nuisance in many ways even if they do not result in direct adverse economics or health effects.

STATE OF AIR POLLUTION

In Mexico city, just breathing is equivalent of smoking two packs of cigarettes a day. Some three million cars and 30,000 factories spew 20 tons of lead, carbon monoxide and other contaminants into the atmosphere of this city of 18 million making it one of the most polluted places on earth. The government has enforced emergency measures including regulating factory hours and keeping cars away from city centre. But that has not prevented thousands of residents to suffer from stinging eyes, sore throats and severe headaches with doctors warning of more severe long term consequences, including lung diseases.

Indian cities are not as yet so badly affected, though in the count of suspended particulate matter (SPM), Calcutta has climbed to the top spot among the major metros of the world.

Air Pollution

Sl. No.	*Pollutant*	*Source*	*Effects*
1.	Carbondioxide	Fuel combustion from heating, transport, energy production	May increase the earth's temperature
2.	Carbonmonoxide	Incomplete fuel combustion in motor vehicles	Deprives tissues oxygen
3.	Sulphur dioxide	Burning of coal and oil	Causes suffocation, throat and eye irritation, produces acid rain reduces crop yields, acidification of lakes
4.	Oxides of nitrogen	Fuel combustion in motor vehicles and furnaces, forest fires	Cause acute respiratory infections, produce brown haze in city and corrosion
5.	Suspended Particulate matter	Smoke from domestic, industrial and vericular sources	Reduces visibility and increases corrosion
6.	Hydrocarbons	Burning of fuels containing carbon like fire wood	React with other pollutants to produce ethylene which is harmful to plants. Aerosol particles reduce visibility May produce unpleasant odours
7.	Ozone	Emissions from motor vehicles	Causes eye irritation, carrodes materials, reduces visibility and is most damaging to plants.

Source : "What is Environment All About ?"
Published by Centre for Environment concerns, Secundarabad (A.P.)

A major source of air pollution in metropolitans is vehicular emissions. All the two and three wheelers in the country run on two stroke engines that burn a mixture of petrol and oil. The combustion is never complete and all of them, therefore, emit a cloud of blue smoke. Our buses and trucks are now powered almost exclusively by diesel engines, and in theory at least the exhaust from them should be far less toxic than that from petrol burning vehicles. By the end of the decade, India should be producing 2,00,000 cars a year. Permission for the manufacture of 3.5 million two-wheelers by 1980 has already been granted. Naturally a direct consequence of this surge is going to be an increase in air pollution.

Prof. J. M. Dave, Dean of the School of Environmental Sciences at JNU and his team of researchers observed that in Delhi 400 tonnes of pollutants are emitted everyday by nearly 5,00,000 vehicles amounting to 34 per cent of the smoke and dust emitted in the city.

The team also pointed out the pollution generated by the different vehicle categories.

"Heavy vehicles—trucks and buses, mostly diesel-constitute only 8 per cent of the total number but consume 70 per cent of petroleum products and produce 49 per cent of pollutants. Two-wheeler and three-wheeler vehicles constitute 63 per cent of the number, consume 9 per cent of petroleum products, and produce 13 per cent of the total pollution. However, scooters and motor-cycles emit twice as much hydrocarbons as cars and the same as heavy vehicles. In other words, they contribute 29 per cent of all the hydrocarbons eight times more per unit of fuel consumed than heavy vehicles."

The major constituents of vehicular emission are carbon monoxide, lead, hydrocarbons, oxides of Nitrogen and smoke.

According to a report prepared by the Institute of petroleum, Dehradun, if vehicle pollution goes unchecked, by 1991-92 the vehicles in Bombay will release 1,07,000 tonnes of carbon mono-xide and 37,000 tonnes of hydrocarbons per year in the atmosphere and in 2,000 A.D. 1,85,000 tonnes of carbon monoxide and 6,93,000, tonnes of hydrocarbons will be released in Bombay city's atmosphere.

All north Indian cities experience the thermal inversion that makes the air nearest the ground colder than the air higher up. This makes the noxious gases, that have been driven up by the heating of the earth in the day time, descends once again at night. What descends, however, is not the emission of those night time hours alone but of the entire day.

The quantity of sulpher dioxide released into air has tripled in 15 year. In industrialised areas Bombay, Delhi and Pune for example, acid rain, the scourge of forests and lakes as in Europe and North America, are now being found. In the State of India's Environment 1984-85, published by the Centre for Science and Evironment, New Delhi, it is reported "of the 48 thermal power stations, offically surveyed in 1984, 31 had taken no pollution control measures and only 6 had their pollution control equipment functioning properly."

All Indian power houses use, rotten coal, at least in terms of its ash content, and not a single power station has de-sulphuration or -nitrification equipment. Most of them do not even have electrostatic precipitators, their notions of pollution control being confined to building tall chimmey in order to disperse the flue gases over a larger area and thus to distribute it to what the authorities hope is more tolerable level.

The smog is another major problem of the metropolitan areas. Besides being a health hazard, it is sometimes so thick that visibility is reduced to such a level that airports have to be temporarily closed. Even the vehicular traffic moves very slowly. Smog being a combination of water droplets of fog, smoke and fumes blenched out into the air form factories, homes and automobiles, these harmful fumes sometimes do not rise high into the air. Instead, a layer of warm air is formed over the cold smog and covers it like a lid.

Now the Country has learnt that further neglect of the environmental aspect is going to be catastrophic. The worst industrial accident on record in India was December 1984, when a deadly cloud of methyl isocyante escaped from a union carbide pesticide plant in Bhopal, in Madhya Pradesh, Killing over 2,000 people and severely disabbling and injuring thousands.

Even, otherwise, the environment is today endangered from industrial pollutants. A case in point here being the Taj Mahal which is being threatened by industrial enterprises coming up in Agra-Mathura region. With increasing congestion the roads and a host of factories blenching smoke, air pollution too is beginning to reach choking levels. In Calcutta, in areas like Dalhousie, Kalighat and Bhowanipur, air pollution is three times the tolerance limit. In Delhi in winter season, it is not mist that blocks the sunlight, but a thick dense smog that settles down for hours.

PREVENTION AND CONTROL OF AIR POLLUTION

Air pollution constitutes a grave danger for the healthy sustenance of all forms of life on this earth, the foremost thing which deserves to be done is to educate the people that the atmosphere is not meant to be dump for all kinds of pollutants. Importance of preserving the health and welfare of man, protection of plant and animal life, prevention of damage to property, ensuring visibility for safe air and ground transport, and maintenance of cleaner atmospheric environment should also be explained. This theoretical effort has to be accompanied by practical measures to control air pollution.

First of all, desirable and harmless air quality standards must be established. Once done, adequate legislation should compel control of pollutants at their sources. Source control in the pure sense deals with the elimination before or during ultimate consumption of potential air contaminant contained in raw materials. Unfortunately lack of technical knowledge and economic incentive prohibits full practice of this classical method. Source control and abatement of formed contaminants are complementary practices in the campaign against air pollution. Source control prevents the emission of contaminants to the atmosphere; abatement renders the emission of contaminants to the atmosphere; harmless and inoffensive. The two controls overlap at many points. For example, in the coal and petroleum industry as much sulphur as is feasible can be converted to hydrogen sulphide in some processing operation and used to manufacture sulphuric acid this is termed as source control. The remaining sulphur can be burned and the resulting mixture of sulphur dioxide and flue gases scrubbed and discharged at high velocity through a tall stack—that is termed abatement, and the net result, a decrease in air pollution. Control of air pollution sources is usually most easily attained in the development stage of a new project and in the architecture design phase of completely new plant. Stringent measures for purification of emissions before they are reduced into the atmosphere should be enforced by law so as not to deviate from the established standards of air quality. Imposition of pollution tax could greatly prevent the defilement of environment. But before undertaking any master plan for air pollution control, it is necessary to know the degree and extent of pollution and the sources responsible for it.

Some higher vascular plants and many non-vascular plants such as mossess and lichnes respond to gaseous and particulate air pollutants at concentrations much lower than those that elicit responce in animals and human beings. These plants are used as indicators of pollutants in air and to monitor their concentrations (biomonitors). In fact, plant biomonitoring data can be profitably used in relation to air quality standards. Closely planted woody plants growing along the roads help in reducing the contamination of lead from automobile exhaust. These plants act as screen and prevent the speed of lead-loaded exhaust from automobiles and also contain the excessive deposition of lead in the soil near highways. Certain weeds such as water hyacinth can check the heavy metal contamination caused by industrial effluents. The stem and leaves of this weed have been shown to selectively absorb and retain heavy metal such as lead, cadmium and chromium from industrial sullage and thus help in reducing contamination.

Technology for control of emission requires to be adopted for all types of vehicle using petrol or diesel and servere punishment should be specified to the defaulters. Bombay Motor Vehicles (Amendment) Act, 1984 empowers authority to suspend, after a grace period of 14 days, the registration of vehicles emitting pollutants beyond a certain specified limit.

Pollution caused by two-stroke engines can only be got rid of by phasing them out. This can be done by setting a deadline by when all manufacturers of motorcycles and scooters will have to obtain the technology for four stroke engines. In case of three wheelers, the government can insist that their engines be designed to run on liquified petroleum gas (LPG). In Bangkok three-wheelers are obliged to have four stroke engines or to run on gas, and in South Korea more and more taxis are running on LPG. What is possible in these developing countries is surely in India.

If manufacturers are given a reasonably long time to change their technology, they can fit this in with their renovation and modernisation plans at no extra cost to themselves, but if the government wishes to shorten the time taken by the change over, it will do well to offer fiscal incentives.

In the same way the surest and easiest way to reduce the pollution caused by burning petrol is to instruct the refineries to produce unleaded petrol only. The octane rating of Indian gasoline is so low that this will pose few problems to either the refineries or to the present generation of cars, all of which have low-compression cylinder heads. As for imported cars the government can insist that only those equipped with catalytic converters or designed to run on unleaded petrol will be allowed into the country. The most intractable problem will be to reduce the pollution caused by diesel burning buses and trucks. If all of them were properly maintained and not overloaded, this would solve it self. But that not practically possible. Thus, the only long term solution is to find a substitute to diesel. This is in any case highly desirable, for the demand for diesel is going up by leaps and bound and at least for the present it cannot be easily replaced in tractors and locomotives.

Research carried out over the past decade or more has shown that with only minor modifications a diesel engine can run on a mixed fuel containing up to forty per cent methanol. Since methanol is a completely clean burning fuel that produces only steam and water and since its flame speed is higher than that of even gasoline.

To curb the other main source of air pollution *i.e.*, the gases of the power stations, complete flue gas treatment plant should be installed. The experience of European power stations shows that such equipment will increase the capital cost of the plant by around 22 per cent.

The percentage of SPM concentration in our country is highest in Calcutta. This has focused international interest on the city's pollution with the World Health Organinsations Global Environmental Monitoring System (GEMS) choosing Calcutta as one of its 'comparison stations'. A continuous monitoring of SPM, as also Sulphur dioxide and oxides of Nitrogen Concentration in three areas of the city *i.e.*, Cossipur, Dalhousie Square and Bhowanipur, is carried out by the National Environmental Engineering Research Institute (NEERI) under the GEMS programme and reports sent every quarter to the WHO headquarters at Geneva. Two other Indian cities have recently been added to the GEMS list of 'comparison stations' Bombay and Nagpur, where NEERI is situated and which among the least polluted metropolitan cities in the country.

If municipal incinerators are operated at a high enough temperature to completely oxidise the organic wastes, odourless air can be obtained. Occasionally, pollution control devices such as settling chambers, bag filters, wet collectors, electrostatic precipitators, incinerators, gas scrubbers, absorbers, catalytic combustion agent need to be used. Construction of tall chimneys for vertical dispersion of pollutants is another alernative. More trees should be planted and looked after because they purify the air.

3

LAND POLLUTION

The area of earth which is capable of supporting life is represented by a thin mentle and there is a very complex relationship between this land and the other components of the environment. There was a time when men and other animals exhausted the resources of a given area. The natural forces could not maintain the balance between the materials consumed and return to soil. They had, therefore, to move to more prosperous regions. However, in modern times, such movements are restricted. Instead, the depleted sources are now replenished by putting in the lost components. In an attempt to do so, man has been responsible for disturbing the natural environment of the soil in various ways, through his domestic, agricultural and industrial activities which have directly or indirectly polluted the soil.

Technology which had seemed to be a boon to the world is a kind of paradox. It provides many of the things which improve the quality of life but in some cases it is responsible for lowering the quality of life, but in some cases it is responsible for lowering the quality of an environment. Irrigation is a blessing when it makes the arid land productive but it becomes a curse when it results in making the soil saline and alkaline. Likewise fertilizers and pesticides are most important, factors for increasing population, but their continuous use particularly of the pesticides is fraught with danger. Under pressure of population which is increasing in an exponential manner, man is forced to use the land more extensively as well as intensively and employ more technology and industrial inputs, which is not an unmixed blessing, as numerous problems of land degradation also to appear.

Soil erosion results in huge loss of nutrients in suspension which are removed away from one place to another thus causing depletion or enrichments of nutrients. Soil erosion is the destruction of soil and the process of removal of the destructed soil either by moving water or wind is called soil erosion. It is estimated that formation of 2.5 cm. soil layer take about one thousand years.

Chief Soil Contaminants and Their Influence

(*a*) Areas around smelting and mining complexes are usually soiled by metals such as cadmium, zinc, lead, copper, arsenic and nickle. These are not only phytotoxic even in small quantities but also render plants unsafe for human and animal consumption. Zinc, often with cadmium, is released into the environment during the use or breakdown of lubricating oils, vehicle tyres, galvanized metals and fertilizers.

(*b*) The major sources of land pollution are the industries such as pulp and paper mills, oil refineries, power and heating plants, chemicals and fertilizer manufactures, iron and steel plants, plastic and rubber producing complexes and so on. Tonnes and tonnes of solid wastes are either dumped or burnt or emptied into rivers. Most industrial furnances produce a grey, powder residue of unburnt material known as fly ash, an important pollutant, besides huge mounds of solid wastes.

(*c*) Modern agriculture has heavily involved in polluting soil through the non-judicious use of chemical fertilizers, herbicides, insecticides and fumignants. Most of these are stable chemicals and remain in the soil for long periods without degradation, and have cumulative effect. Pollution arising from pesticides usage have been among the greatest causes of concern in the field of agricultural pollution and much has been written on the subject. Pesticides residues are found in soil air and water as well as in living organisms. Apart from killing the living organism present on the surface of the soil, they reach even the deeper layers through tilling and irrigation of the land, killing still more living forms. With the continuous use the soil micro organisms loose their ability of nitrogen fixation.

(*d*) Quite a large amount of solid rubbish is contributed by our households in the form of domestic wastes. Some common examples are groceries, food scraps, vegetable remains, packing materials, paper, remnants of used coal, ash, wood, metals, plastics, ceramics, glass, etc. Many of these are non-reusable. All these go to constitute heaps of municipal refuse. It is not properly disposed off, this can prove perilous. Such places often become a home for rats, flies, bacteria, mosquitoes and a large number of other vectors, having the potential of causing many human diseases.

Prevention and Control of Land Pollution

It should be imperative for the industries to install collectors to remove the particulate waste from the chimneys. Appropriate methods should be developed to dispose off or utilize the other type of pollutants. The garbage, instead of burning in the open, can be used not only to produce energy but also filler for cement, bricks, ashpalt, pavings. Some of these wastes, if properly shifted and separated can even be recycled as raw material for industries.

The sanitary landfill is in fact a better remedy for large objects because it brings about inexperative biodegradation of such trash without causing much pollution, disease, etc. Low-lying watery areas and ditches can also be filled by this stuff and land thus reclaimed used for making gardens, parks, playgrounds or even apartment complexes. Realistic management of domestic wastes by municipalities is a must for proper disposal and economic recycling.

Animals refuse and agricultural waste can be utilized as manure and for the production of biogas that in turn can help to generate electricity. Sludge from biogas plants after proper conditioning can be used as soil builder and fertilizer.

India embarked upon a pilot project to extract combustible gas and eventually electricity from a garbage dump when the Delhi Energy Development Agency under the auspices of the Department of Non-conventional Energy Sources of the Government of India planned of tapping the Sanitary Landfill at Timarpur in Delhi in 1986. The area has a potential of generating 8MW of electricity from landfill which is spread over an area of about 80 acres.

Large quantities of municipal solid wastes are disposed of in landfills which are potential sources of biogas, methane constituting about 60 per cent and the rest of it carbon dioxide. The cost of generating at the project works out at 10 paise per unit compared to 62 paise per unit through conventional sources of electricity generation. The experts feel that one well is enough to extract gas from one acre of landfill and well has a capacity to produce 100 KW and from 10 acres, one MW can be produced. At present there are eight wells dug up at a distance of 200 feet from each other.

According to authoritative sources of the Department of Non-conventional Energy sources, to extract gas in enough quantity there should be properly constructed landfills with a height of at least 40 feet, refuse should be compacted and covered by earth from layer to layer and landfilling should be uniform. The gas can be extracted by drilling perforated pipes into the landfill, piping them together and pumping the gas to a common plant through collecting pipes. This gas can be utilised for power generation through IC engines or gas turbine or a cooking fuels after cleaning.

The department of Non-conventional Energy Sources has taken up another pilot project where advanced technology is being tried to dispose of municipal solid waste by incinerator has an inbuilt precipitator to take care of air pollution and uses 300 tonnes of garbage per day producing 3.75 MW of power. The engineering work here has been entrusted to Bharat Heavy Electricals with Volound Miljoteknik, a Danish firm.

The specialised equipment not produced in India is being imported from Denmark. If the project succeeds, the Department of Energy may think of having bigger incinerators at different places in the city so that the cost of carrying garbage and power can be kept low.

Making biogas from the garbage will also diminish land pollution to some extent. A break-through has recently been made experimently in a joint venture launched by the Burn Standard Company Ltd., Central Mechnical Engineering Research Institute (CMERI) of the CSIR and the Calcutta Municipal Corporation. This works on the conception that any organic matter on biodegradation gives out methane, carbon dioxide and trace of carbon monoxide, hydrogen sulphide and some oxides of sulphur and nitrogen accompanied by obnoxious smell. This is a very well known natural phenomenon brought about by a group of microbes called methanogenic bacteria, but these bacteria act in full form only in anaerobic condition, that is, in the absence of oxygen and at 37 degree centigrade temperature.

In USA, scientists have been able to convert organic waste-rubblish, wood, shaving, rags and even cow manure—into clean, high energy oil or gas. A project sponsored by the environmental protection agency breaks down organic wastes into oil, gas and a combustible hard waste by heating it to temperature in the absence of oxygen. The advantage of this process—called pyrolysis—over the burning of wastes is that it produces a transportable fuel. The U.S. Bureau of mines used and method by which it converts wastes into oil by treatment with Carbon monoxide and water under pressure. By another method they covert manure into a pipline gas by treatment with hydrogen at high temperature. Thus, we see that there is no waste in the nature which cannot be recycled and put to use for our daily benefit.

4

WATER POLLUTION

As per G.M. Fair and J.C. Geyer water pollution may be defined as "introduction into a body of water of substances of such character and in such quantity that its natural quality is so altered as to impair its usefulness or render it offensive to the senses of sight, taste or smell."

It is estimated that man can survive for 20 days without food, but starts struggling for life in the absence of water just after one day. Water is needed for the maintenance of life of plants and animals, for navigation, hydroelectric power and for disposal of sewage. Supplies of fresh water also provide the aquatic organisms with dissolved oxygen and some essential minerals and nutrient making it indeed the most vital element for life. That is why most of our villages, towns and cities sprang up in the past near places where plentiful water was available in the form of lakes and rivers. History tells us that man civilizations perished or migrated to better locations in the absence or scarcity of water. In primitive ages the man used to live near the rivers. The greatest use of water was for irrigation and agricultural persuits, while only a small amount was consumed by the people. The use of water for drinking and cooling was limited to that amount that people could easily take from the wells or the stream in their vessels. The management and conservation of water resources have become important issues,

The most important contributors to pollution of water are sewage, oil, industrial and agricultural wastes. These can be divided as degradable and non-degradable. Degradable pollutants, mostly domestic, sewage, can be rapidly decomposed by natural process.

Non-degradable pollutants (inorganic chemicals such as salts, chlorides, metallic oxides and toxic and other waste producing materials) are those substances in which there are no evolved natural treatment processes that can keep up with the rate of man made input eco-system. These either do not degrade or degrade only very slowly in the natural environment.

Water pollution not only changes the physical properties of water such as colour, odour, turbidity, taste and temperature but also makes it acidic, alkaline or saline due to the presence of dissolved or suspended chemical substances.

Water pollution is caused due to physical, chemical and bacteriological inpurities in water.

1. Physical Impurities

Include turbidity, taste, colour and odour.

(*a*) Turbidity is caused by suspended and colloidal matter.

(*b*) Colour is due to presence of minerological compounds such as iron-oxide.

(*c*) Taste and odour are due to presence in water of organic matter dissolved during passage through the ground or from industrial work, micro-organism such as algaeal growth.

2. Chemical Impurities

Chemical impurities are due to, carbonates and bicarbonates of calcium and Magnesium, sulphates and chlvrides of calcium and Magnesium and carbonates—bicarbonates of sodium. Nitrates, chlorides and Fluorides of Sodium, Iron oxide and Manganese,

They will create turbidity, hardness and alkalinity, bad taste and odour problems.

3. Bacteriological Impurities

Due to pathogenic bacteria, bacteriological impurities arise in water. Their presence is noted if E-Coili bacteria are present.

So the bacteriological analysis involves the following tests :

(*a*) Stand plate count;

(*b*) E-Coli test.

SOURCES OF WATER POLLUTION

There are four primary sources of water pollution :

(*a*) Urban

(*b*) Industrial

(*c*) Agricultural

(*d*) Natural

The urban sources are of two types :

(*i*) *Controllable Source*

Normally the sewerage system serves the business and commercial areas, the residential districts and industrial area. The extent to which this system services the city and its surrounding suburban and industrial fringe is a measure of controllable source.

The discharge of huge amounts of municipal and household wastes into rivers and canals is one of the major sources of pollution of our water bodies. Most sewerage systems mainly contain human and animal wastes. However, the amounts and types of other kind of refuse from modern living are continuously increasing.

Some of which pose processing problems, for example the synthetic detergents. Now complex compounds such as cleaners and water based paints also find their way into it. Sewage, garbage and organic materials dumped into the water bodies kill the fish because of reduction in oxygen concentration.

Being rich in organic substances, sewage provides nutrition for various decomposer bacteria and fungi. The decomposition products of sewage, detergents and fertilizers-phosphates, nitrates and ammonium compounds—promote the growth and reproduction of algae and results in the formation of algal blooms. These large algal masses decay and cause deoxygeneration of water. Most of the epidemics of big cities are caused through water borne diseases such as typhoid, cholera, dysentry and jaundice. Enhancement of bacterial, viral and other parasitic populations in polluted waters endanger human health.

(b) *Uncontrollable Source*

All urban wastes that reach the stream other than through the organised sewerage system and treatment works are uncontrollable source. This constitutes a great contribution of stream pollution. The uncontrollable source is usually intermittent, associated with the occurrence of rainfall.

2. Industrial Source

The industrial source is divided into :

(*a*) The part that is connected to community system.

(*b*) The part that is dealt with by independent private systems.

Today industry contributes more water pollution than do house-hold users. The major industrial pollution are the chemicals, primary metals, paper and food industries. Wastes from industries such as pulp mills, leather tannaries, sugar and oil refineries, jute mills, coal washeries, petroleum and chemical fertilizer plants, are mostly complex organic compounds. These are emptied directly into natural waterways.

The effluents from industries are resistant to breakdown. They result in disastrous consequences upon the existing eco-systems Rapid industrial development has indeed been responsible for the neumerical decline of macrophytes which consitute an important constitutent of acquatic eco-system. Chemically polluted water either damages the growth of crops or changes the acquatic vegetation due to artificial nutrients, and is totally unfit for live-stock to drink. Cyanides, acids alkalis and other industrial wastes affect the inhabitants of rivers upto several kilometres-downstream.

In our country, there are several examples of pollution in the rivers. The 14 rivers which provide over 85 per cent of the country's drinking water are polluted, according to environmental experts. Not only does this adversely affect the health of so many people, by the resulting water-borne diseases such as hepatities, typhoid and dysentry, but accounting for about two-thirds of all illness.

In the State of India's Environment 1984-85, published by the Centre for Science and Environment, New Delhi, it reported, "No river in India is safe for fish. Some 1,500 industries situated along the Ganga release untreated effluents into the river. Industries on the Yamuna's banks release water at 60°C making it difficult for fish to survive. Almost all chambal tributaries big or small, are highly polluted making chambal's fish catch one of the lowest. Paper, pulp and chemical factories, sugar, cement and other industries are pouring millions of gallons of wastes per day into the river sone. No fish servive upto 22 km downstream of Dehri on son. The Damoder, Hoogly, Bhadra, Godavari, Cauvery are no exceptions to the rule."

Thus polluted rivers, lakes and large dams also seriously affect riverine fish, an important source of protein in India. About 100 million Indians eat fish which constitute 2.3 per cent of protein supply, and 22.4 per cent of total animal protein supply.

Thousands of fish perished in Gomti river near Lucknow in April 1984 due to effluents. In Kerala, because of the indiscriminate use of back water and lagoons by the copra industries as dumping grounds for coconut husks, one finds that these waters slowly dying. The discharging of water that is low in dissolved oxygen, high carbon dioxide and gases such as Hydrogen sulphide by deep reservoirs can kill fish downstream.

3. The Agricultural Source

Agricultural practices such as crop and live stock production are sources of pollution also chemical fertilizers, pesticides and herbicides are of increasing significance at a source of pollution. The rapidly increasing use of inorganic fertilizers, especially, the readily subtle nitrogenous salts, has led to nutrient enrichment of many of our water bodies. This kind of agriculture drainge encourages algal growths and contaminates drinking water particularly with nitrates. Pesticides and fungicides from agricultural run off when taken in by fish and other aquatic organisms, become concentrated in the bodies of organisms, higher in the foodfeb. For example the concentration of DDT may be very low in river water. But some fish in the river contain such high proportions of DDT that they become quite unfit for human consumption.

4. The Natural Source

The natural sources are storm, wash, seepage from groundwater, swamp drainage and aquatic life of the stream. Natural rain water has an approxmiate pH of 5.6. However, during the last 25 years reports from many developed countries have indicated that rain water has much higher acidity. This is brought about by strong acids produced from industrial pollutants such as sulphuric acid formed from sulphur dioxide and nitric acid from nitrogen oxides, and probably hydrofluoric acid from fluoride. These acidic gases travel long distances. Some of these pollutants over the city ascend skywards and then travel down through precipitation as rain, dew and snow. Acid rain has eliminated fish in hundreds and thousands of lakes in USA, Canada and Scandinavia. The portion of rain water that flows over the surface and called run-off pick up organic and suspended matter, whereas the portion percolating through the ground has got minerological, organic and inorganic matter which it gathers while traversing through the underground strata before reaching the waterstable.

The last 25 years have witnessed several instances of coastal water pollution. The main source of pollution is oil. This oil pollution is due to the wrecking of oil tankers and accidental oil spills. This constitutes a major threat to the oceanic eco-systems.

There is also now-a-days thermal pollution of water. With the increased utilization of atomic energy as a major source of power, the problem of thermal pollution is assuming dangerous proportions. The production of electricity by nuclear power requires billions of gallons of cold water to remove waste heat. The warm water emptied into the waterways adversely affects the aquatic organisms. The entire aquatic eco-system is affected by changes in temperature that disrupt the food chain and upset the entire balance among the living organisms. Radio active pollution of water is another threat to cities where nuclear power plants are located.

WATER STANDARDS FOR DIFFERENT USES

1. Quality of Water for Drinking

Supplies should be drawn from the best available source. If the source cannot be adequately protected against pollution the water must be treated to ensure its safety. Possible hazards should be identified by sanitary surveys and eliminated.

BACTERIOLOGICAL QUALITY

The coliform index is a measures of the concentration of coliform organism or E-Coli in a water sample. It is defined as the reciprocal of the smallest quantity of sample (in ml) which would give a positive E-coli test. This index is now obsolete and now Most Probable Number (MPN) is the one commonly used. MPN is defined as that bacterial density which if it had been actually present in the sample under examination, would more frequently than anyother, have given the observed analytical result.

PHYSICAL CHARACTERISTICS

The physical characteristics of water should be examined at least once a week, sample being drawn from representative points through out should not be so high as to offend the senses of the sight, taste or smell of the consumer. Maximum acceptable values are :—

Turbidity — 5 units

Colours — 15 units

Odour No. — 3

CHEMICAL CHARACTERISTICS

Drinking water should not contain impurities in hazardous concentrations be excessively corrosive or retain treatment substance in excessive concentrations.

QUALITY OF WATER FOR INDUSTRIAL USE

Absence of odours, slick and Unslightly suspended or floating matter, D.O, *i.e.*, Dissolved oxygen (near saturation) mg/l as daily average, always 1.0 mg/l, pH 5.0-9.0, temperature 21°C, dissolved solid 750 mg/l as monthly average, always 1000 mg/l.

QUALITY OF WATER FOR RECREATIONAL BOATING

Absence of slick, odours, any visible floating and suspended solids, DO=5.0 mg/l during at least 16 hr/day always 3.0 mg/l, Co_2 40 mg/l, preferable less than 20 mg/l, pH 5.0-9.0, daily average preferable 6.5.—8.5, temperature 20°C, toxic substance 0.1 median 48 hr—tolerance.

QUALITY OF WATER FOR BATHING

No visible sewage matter, DO near saturation; bacterial standard coliform groups per 100 ml. A free chlorine residual of at least 0.4 ppm must be maintained throughout the bathing pool.

PREVENTION AND CONTROL OF WATER POLLUTION

It is essential to have sewage water treatment plants for every town and city so that the biodegradable and non-degradable pollutants can be removed from it and pure water obtained by recirculation. Modern techniques have certainly made it possible to achieve this.

SEWAGE WATER TREATMENT

Sewage water can be disposed of either by land treatment of dilution. Waste water consists of large amount of waste products and disease bacteria. If sewage or waste water is directly disposed off either by dilution or land treatment, natural force of purifcation in the form of air, wind, sunlight, bacteria and other organisms convert it into harmless substances. As far as the amount of waste water is small as compared to land available for land treatment or in relation to the quantity of dilution water there is no problem of disposal. But when sewage to be disposed off is large enough in relation to land available for land treatment or in relation to the quantity of dilution water, there is difficulty of disposal. In such case the waste water in case of dilution will pollute the stream. And in case of land treatment septic condition may be developed and land may become sewage sick. So the waste water is given some treatment so that it may be safely accepted by the land or receiving water.

Generally the sewage is first passed through a series of screens to remove large objects, and then through a grinding mechanism to reduce the remaining objects to a small size so that they can be effectively handled. The next step is designed to remove heavy grit and other suspended solids through several settling chambers. Upto this stage; the process is called primary treatment which is rather inexpensive. But the water still contains a large concentration of pathogenic and non-pathogenic organisms and ample quantities of oranic matter. This now acquires an accelerated microbial action for further decomposition. One commonly used method is to distribute polluted water through continous sprays over a bed of stones so that the growth of several forms of life requiring nutrient and oxygen is encouraged. Gradually a fast moving food chain is set in operation. Food chain is the pattern of flow of energy in a natural eco-system. Bacteria consume organic matter, portozoa thrive on bacteria and worms, snails, insects come next. Every form of life present there (including algae) plays its role in converting the high energy chemicals into low energy types. Even after this biological (secondary) treatment, the water is not yet fit for drinking.

The micro organisms need to be killed, and several other chemicals present in it require to be neutralized. The final step *i.e.*, testing treatment is, therefore, a disinfection process, usually chlorination. Additionally, flocculation helps to remove lime, alum and some salt of iron. Tertiary treatment also involves diverse operations for the removal of nitrogen and phosphorus. These general principles of purification of sewage water are modified according to the origin and nature of pollutants discharged into it. Alternatively a modern method of sewage water treatment is by aeration. Sewage can also be pumped into aeration tanks where it is mixed with air and bacteria ladden sludge so as to set up a similar food chain.

Constant research in processing sewage is required to keep up with the many new products entering the market and finding their way into the disposal systems. The increase in population and urbainzation place an ever increasing burden upon sewage system. Many countries have devised various methods of recycling sewage water. Ten years ago USA invented a machine which turn waste water and sewage to pure water more cheaply and efficiently than other method. Raw sewage from the waste treatment plant's tank is fed through a sequence of electrically driven spinning devices. On each of the discs grows a colony of different biological organism which feed on different water impurities. At the end of the process, the combination of organisms draws off virtually all impurities from the water. As a final filtration, the water passes throw an activate carbon filter that removes any remaining traces of organic matter and colour.

In India the first ever experimental project using atomic radiation process for the treatment of Sewage water has recently been set up in Baroda by the Bhaba Atomic Research Centre. The project has come up on a 830 sq. m. plot of land donated to the BARC by the Baroda Municipal Corporation near the former's existing conventional sewage plant. The project costing Rs. 75 lakh uses cobalt—60 atomic radiation process to rid of the conventionally treated sewage water of any residue of harmful bacteria content before it is recycled for an alternative use. The Baroda project has been set up to test the efficacy of atomic radiation as effective measure of water pollution control.

Currently, water hyacnith has come into prominence for purifying domestic and industrial waste waters. The plant regenerates rapidly and has a tremendous capacity to accumulate heavy and even radioactive metals. It is efficient in absorbing nitrogen, phosphorus and similar chemical pollutants. The polluted water fed into reservoirs or lagoons with water hyacinth, becomes clean and free from 75-90 per cent of its pollutants. Besides, this plant has been used as a new source of food, fertilizers, energy.

Synthetic herbicides and pesticides pose a challenge to the natural degradative process. Therefore, judicious, efficient and opitum use of organic manures requires to be encourgaged. The use of microbes for the breakdown of these synthetic compounds is another answer to this problem. For instance, specific micro organisms are being evolved by molecular breeding process to convert the chief ingredients of herbicide agents, 2, 4, 5-T, into Carbon dioxide and chloride.

Microbiologists have also found a solution to the difficulties of treating waste water polluted wlth cyanide and heavy metals through bacteria wihch can only withstand but are also able to grow at high levels of cyanides. Millions of these bacteria are introduced in each of the several rotating discs of the plant's main processing units. The sticky body surfaces of these bacteria pick up zinc, iron and other metals from the water as it passes over the plates. They also ingest the cyanide so toxic to the fish and marine life. This water is subsequently passed through clarifier and filters and then released into the streams.

GANGA ACTION PLAN

Realising the devastation caused to the population of the cities and towns of States and the Union Territory located along the Ganga, through its run of 2,525 kilometre from Gangotri

(Himalayas) to Ganga Sagar (Bay of Bengal), the Government of India has launched Ganga Action Plan to restore its water quality. Purification work of Ganga is in full swing. Cleaning of Ganga at Hardwar, Varanasi, Allahabad and Kanpur is being done with great enthusiasm. This tremendous task of cleaning up the water of this river to a desirable quality is likely to take several years.

> "The Ganga Action Plan is not just a Government Plan it is a plan for all the people of India, one in which they can come forward and participate."
>
> —Rajiv Gandhi

Ganga Action Plan covers 27 cities along the river, work is in progress in most of the towns. In Varanasi, under a multi-sectoral city Development Package, ghats are being renovated, sewers are being intercepted and work is in full swing on two of the three sewage treatment plants. In Kanpur, the sewer network and treatment plants at Jajmau are part of larger package of development schemes. In Allahabad Gaughat pumping station is being expanded. Nallahs discharging waste into Yamuna river are being intercepted, work is also proceeding fast at the Naini Sewage farm.

The major aspects of the programme are interception and diversion of sewage now flowing into the river, integrated sewage treatment plants with energy recovery, low cost sanitation, etc.

For the first time in the country a wide range of schemes have been taken up to deal with domestic wastes and industrial effluents polluting the river. The Ganga Project Directorate monitors closely all major schemes costing Rs. 50 lakhs or more. State level specialised agencies are responsible for execution and maintenance. Many schemes are expected to be completed by the end of the Seventh Plan.

The recovery of resources from the treatment of wastes, monitoring of water quality and conservation of the fish and aquatic life in the river are important aspects of Ganga Action Plan. All the major universities along the river and scientific organisation under the council of Scientific and Industrial Research and ICAR network are involved in the formulation of the schemes that will help us to know the river better.

The purity of a river water is usually measuread by oxygen standards. One is Dissolved oxygen (DO), the more the oxygen the better the quality. The second is Biochemical Oxygen Demand (BOD) which is a measure of organic matter present in water, the less the BOD, the better. Another standard is coliform, the less is its count, the better.

Scientists have developed a composite Index integrating the above three parameters. The index varies according to the water use. For bathing, for which Ganga is widely used, the index number is 50. The river quality varies from season to season because of the changes in the flow of water. So, to assess quality of water, samples are taken in different seasons.

Arrangements have been made for regularly monitoring water quality of the river at various locations and at different times. The water quality is reasonably good in Hardwar-Rishikesh. But it is below standard in other places. The main thrust of the Ganga Action Plan is to prevent pollution so that the purity of the river can be restored all along its course. German scientists have developed a new way to purify water without adding any chemical substance as reported by United Nations Industrial Development Organisation (UNIDO). This method is based on 'anodic oxidation'. All dangerous bacteria and viruses are eliminated and water becomes potable.

In comparison to other known methods like treatment with ultraviolet radiation, ozone or chlorination the new method builts its oxidation agent out of water. The new system is simple and safe for sterlization of drinking water, industrial water and surface water. This new method is expected to prove useful in achieving the United Nations goal, aimed at providing both urban and rural population with clean drinking water, between 1981-1990.

MARINE POLLUTION

The coastal waters of India are extremely rich in food and mineral resources. The production of this environment, therefore, is the basic need at present. There are a number of factors responsible for the pollution of coastal waters. These include sewage discharge from cities, industrial wastes, dumping of garbage and agricultural wastes, petroleum, etc. The magnitude of oil pollution is increasing rapidly, particularly in coastal waters. Approximately two-thirds of the world's oil production is transported by sea. About 0.1 per cent of this spills in sea. Oil pollution arises from tanker accidents, deballasting operations and tank washing, refinery effluents, losses from pipelines and off shore production platform. The input of petroleum and of petroleum products to marine environment from different source have been estimated to be quite considerable.

In the early instances of oil spills in water bodies, detergents were used to remove them from the water surface. But these provided to be even more toxic to aquatic life. Recently a product bregoil' resembling saw dust from paper industry waste has been employed. It rapidly absorbs oil. The residue can be conveninetly handled and even burnt as a fuel. Bregoil may make the cleaning of world's growing number of oil spills much easier.

Monitoring and research in the area of marine pollution control are fast developing activities in India. The studies on marine environment have been undertaken since mid-seventies by the various institutions in the country. Surveys for the collection of baseline data on all the potential pollutants in the marine environment are in progress and a good data base has been built for this purpose. Data are also collected of the concentrations of heavy metals dissolved from and in zooplankton and fishes collected from the Arabian Sea and Bay of Bengal. Some analyses of DDT and its metabolics has been carried out. In most cases the value have been found to be below the detection limit. The department of ocean development is entrusted with the responsibility to look into various aspects of marine pollution. The department will shortly be establishing a network of stations alongwith the coastline for monitoring pollution and other ocean parameters along the Indian coasts and in its islands.

According to a recent decision taken by the Centre every major port in India will have to create a separate anti-pollution cell to control oil and other pollutions. To began with Bombay, Calcutta, Madras and Visakhapatnam ports will have cells. In other ports also one of the officers of the Marine Department will be incharge of the anti-pollution work. Pollution in areas beyond the control of the ports would be taken care or by organisations like Coastal Guards. The International Marine Pollution Convention (MARPOL) stipulates that ships and ports should have certain facilities for reducing pollution. Steps are being taken to introduce anti-pollutant measures in ports.

5

NOISE POLLUTION

All cities are noisy places. Noise pollution imbibes in itself a serious threat to the quality of man's environment.

Noise may be defined as 'unwanted sound' and noise pollution as unwanted sound dumped into the atmosphere without regard to the adverse effects it may have. The term 'noise' in the electronic communication system is also referred to as perturbations that interfere with communications. Such noise increases with complexity and information content of systems of all kinds. Thus, man faces a growing problem with electronic pollution as radio communication intensifies.

Therefore, in the broadest sense noise pollution can be stated to be an 'unforeseen backlash' in concentrated use of power.

It is the loudness and duration of noise which is disturbing, and causes physical discomfort and temporary or permanent damage to hearing. The unit of measurement of intensity of sound is called the decibel abbreviated as dB. Human ear is sensitive to an extremely wide range intensity from 0 to 180 dB, 0dB being the threshold of hearing, whereas 140 dB marks the threshold of pain. Threshold means the lowest intensity at which stimulus becomes perceptible.

Ordinary conversation which is in the frequency range between 30 and 60 dB, while noise under a jet plane at take-off may rise in excess of 160 dB. The effect on man varies with the frequency or pitch of the sound. The sound pressure level is judged to be of greater loudness for higher pitched than for the lower pitched sounds.

Steadly exposure to sound at level 90 dB is believed to cause a loss of hearing. Sound must be considered on potentially serious pollutant and grave threat to environment. Therefore, measurement, abatement, regulations and legal restrictions on noise pollution must be considered on a high note. Noise abatement would make a good crusade for the younger coming generations, as an environment free of unwanted noise is likely to be a quality environment.

In Bombay, a high court committee to investigate noise levels found that in most parts of the city they ranged between 57 and 91 decibels, far higher than the limit of 55 decibels fixed by the World Health Organisation.

Health Hazards of Noise Pollution

Noise not only interferes with communications but also affects our peace of mind, health and behaviour. Sudden loud note can cause acute damage to the eardrum and the tiny hair cells in the internal ear whereas prolonged noise results in temporary loss of hearing or even permanent impairment. It causes headaches and irritability. It affects the sensory and nervous systems producing several related physical ailments.

Sources of Noise Pollution

While the modern household gadgets such as mixer-*cum*-grinder, vacuum clearner, washing machine, coolers, air conditioners greatly enhance the levels of sound and are potentially dangerous to health, loud speakers, not only disturb the students of their studies, but also the peace of the locality. Similarily loudly played radios, stereos, televisions are other major sources of noise pollution. Printing presses and small-scale industries as well as trucks, autos, motorcycles, aeroplanes of all types, contribute to the noise problem in almost all the larger cities of India.

Noise Pollution Control

In the electronic age it is almost impossible to completely get-rid of the malady of noise pollution. However, there can be ways and means to reduce noise intensity so that its influence can be minimised.

The means of noise control are :

(*a*) to manipulate the source so as to reduce the noise at its origin;

(*b*) to interrupt the path of transmission;

(*c*) to protect the receipient.

Legislation and public policy are essential. Noise could not be completely diminished but it can be minimised to a great extent.

To minimise noise following measure should be taken :

(*a*) Nobody should be permitted to create noise in silent zones.

(*b*) Noise producing vehicles should use silencers, and if their silencers are not working they should not be allowed to ply on the roads.

(*c*) Standards for noise control measures should be set up for industry and community.

(*d*) Building codes should enforce sound proofing in the canstruction of buildings.

(*e*) Plants are efficient absorbers of noise, especially high frequency. In metropolitan cities green belt vegetation may have a great value in minimising sound, (*f*) Personal protection can be achieved by holding our hands over ears under noisy situations, or by running away from the source, or by simply stuffing in the ears bit of cotton through it does very little good. Use of ear plugs can also reduce noise almost by 40-50 dB.

The safe limit of noise, is 90dB. The existing ordinances, Acts, against noise pollution should be revised from time to time depending upon the changing nature of the sources.

6

RADIATION POLLUTION

Radiation pollution is caused by radioactive substances. Radiation is a physical phenomenon in which energy travels through space. Radiations are divided into two main groups.

(*a*) Non-ionizing *e.g.*, ultraviolet rays,

(*b*) Ionizing *e.g.*, x-rays, alpha, beta, gama rays, protons and neutrons.

Sun is the source of spectrum of radiations such as radio-waves, intra-red, ultraviolet, x-rays, gama and cosmic rays. In addition, radioactive isotopes give off subatomic particles such as protons neutrons, electrons and helium nuclei called x-particles in the process of decomposition from an unstable state to an stable state to a more stable condition.

Non-ionizing Radiations

Radiations of shorter wavelength which have greater energy may be harmful to micro-organisms but are capable of injuring only the surface tissues of higher plants and animals. They also increase the rate of mutations. Some plants though absorb maximum radiation in the ultraviolet region, they can grow well in green houses where practically all Ultra violet ray is filtered out. This indicates that Ultra violet ray is not essential for plant growth.

Biological systems possess certain enzymes which enable the cells to repair or eliminate the genetic damage. Ultra violet induced damages are repaired by activation of an enzyme upon exposure of dimer—containing DNA to strong illumination. They can also be repaired in dark which involves removal of a part of DNA carrying the dimers with the help of the activity of enzymes.

Ionizing Radiations

Ionizing radiations are very high energy radiations that are able to remove electrons from atoms and attach them to other atoms thereby producing positive and negative ion pairs are known as ionizing radiations.

Ionizing radiations induce mutations and breaks in chromosmes. The damage is more during the cell division process. In man, the senistive areas are epithelial linings of the skin and intestine, blood-forming cells in the bonemarrow, and reproductive cells. Excessive use of x-rays causes death of tissues. Cosmic rays, radiation from rocks and soil in the form of radioisotopes and radioactive compounds reach plants which are then transferred through them to animals. The damage due to radiations is dependent upon developmental stage of an organ, age, sex of the organism.

Man-Made Radiations

The greatest exposure to human beings comes from the diagonistic use of x-rays, about 70 millirems a year per person; and radioactive isotopes used as tracers. Radium dial Wrist watches and TV sets add another millirem per day. Besides, we are increasingly being exposed to cosmic radiations because of more deforestation. Cosmic rays are radiations from outer space that are mixture of corpuscular and electromagnetic components. The intensity of cosmic rays in the Biosphere is low and they are a major hazard in space travel. Radioactive wastes given off by nuclear generators also constitute a potential radiation hazard.

Thus, on an average each one of us is exposed to 200 millirems of radiations annually.

Nuclear Wastes

Nuclear Wastes from atomic power plants come in the form of spent fuel rods of uranium and the by-prodcuts such as plutonium. It is estimated that these can remain toxic to humans for over 200,000 years. Radioactive iodine, another waste product from power plants, can cause cancer of the thyroid gland. Waste coming from the production of nuclear weapons produces radioactive strontium and caesium, both of which are carcenogenic. These materials generate heat and penetrating radiations from centuries. The nuclear wastes such as contaminated dust, debris, clothing, industrial clothing trash, etc. when dumped into the disposal sites can defile the atmosphere and even seep into the soil and pollute the drinking water. It is, therefore, necessary, that the wastes coming from power stations and defence establishments should be carefully handled, isolated, burried and protected.

Nuclear Fall-Out

The fall out from atomic explosion also cause pollution in the atmosphere. The radioactive dust that falls to earth after atomic explosion is called radioactive fall-out. These materials mix and interact with natural particulate materials in the atmosphere and increasing amount of man-made air pollution. The kind of radioactive fall-out depends upon the type of bomb.

Fall out from weapons differs from atomic waste materials in that the radio-nuclides are fused with non-silca, and dust and whatever happenes to be in the vicinity to form relatively insoluble particles. The smaller particles adhere tightly to the leaves of the plants where they may not only produce radiation damage to plant's tissue but may be injected by grazing animals and then to their alimentary canal.

The fall-out can enter the food chain directly at the harbivore or primary consumer. In general the total amount of radioactivity decreases with distance from a nuclear test. There are certain exceptions to this especially strontium-90 which reached a peak in wild animals populations 50 to 100 miles from the 'ground zero' of the explosion.

The amount of fall-out received in an area is roughly proportional to the rainfall. The quality of fall-out radionuclides that enters food chains and eventually becomes transferred to man depends not only on the amount received from the air but also on the structure of the eco-system and the nature of its bio-geochemical cycles. In general a larger proportion of fall out will enter food chains in nutrient poor environment.

Fallout radionuclides have been and are being passed on to man through the food chain, although concentration in human tissues are not generally high as those in sheep and deer. Man is somewhat protected by the position in the food chain and by food processing and cooking which removes some of the contaminants.

7

ENVIRONMENTAL PROTECTION AND LAWS

Introduction

The environment today is endangered from industrial pollutants; indiscriminate deforestation, vehicular emissions, noise, etc. It is a matter of global concern. Developing countries are affected very much by the environmental pollution. In India government and non-government bodies are emphasizing very much to protect the environment. Our Government has taken up many measures for the prevention and control of environmental pollution.

The Central Government has established a Ministry of Environment in 1985 and made its clearance mandatory for the sanction of new investment projects. It has tightened its pollution control laws for industry, and is exhorting the State Governments to implement these more vigorously. 'Ganga Action Plan' is one of the measures taken by Central Government to purify river Ganga. In February 1985, Prime Minister Rajiv Gandhi formed a Central Ganga Authority'. The chief object of this authority is to make river Ganga free from pollution and to prevent its pollution in future. Under the presidentship of Prime Minister a board has been constituted, consisting of 10 members. Purification of the Ganga river is being carried out at fast speed. The major work is in fullswing at Haridwar, Varanasi, Allahabad and Kanpur.

Recently, it has been proposed to release turtles in the specific areas of the river Ganga as a part of Ganga Action Plan, to help in maintaining its ecological balance. The turtle scheme will cost more than Rs. 30 lakh to be shared by the Union Government and the Uttar Pradesh Government.

The scheme is being implemented only in the Ganga, other rivers are not covered under this scheme. The agency making use of turtles is the wildlife wing of the Forest Department of Uttar Pradesh.

The scheme is for captive breeding of fresh water turtles and releasing them in those parts of the Ganga where their population has decreased due to man's interference and several other reasons.

The Union Government will provide financial help up to Rs. 13.50 lakhs as non-recurring expenses during the Seventh Plan under the Ganga Action Plan. The total financial assistance is likely to be of the order of Rs. 34.50 lakh for this scheme. To contain water pollution in the rivers the steps taken by the Government include basin studies of 14 major rivers for determining water quality, monitoring of water quality at 170 stations, identification of water polluting industries, setting up of minimal national standards for effluent discharge, and regulatory and legal measures under the provisions of relevant acts. Of the 4,054 water polluting industries identified, 2,076 have set up effluent treatment plants. Also, 1,748 prosecutions were launched under the Water (Prevention and Control of Pollution) Act, 1974 out of which the authorities took action as per the court's orders against 369 units.

Upto January 31, 1987 Rs. 54.76 crores has been spent on Ganga Action Plan. In 1988 year's budget, Rs. 57 crore has been kept for the plan.

Ninteen schemes at a cost of Rs. 24 crores in Allahabad and 34 schemes at a cost of Rs. 45.77 crores in Varanasi are being implemented. Till Jan 31, 1988, 15 schemes in Allahabad and 29 schemes in Varanasi at a cost of Rs. 15.92 crores and Rs. 36.23 crores respectively have been sanctioned out of these four schemes—Three in Varanasi and one in Allahabad, have already been completed.

The work for interception and division of waste water will be completed before January 1989, in Allahabad and during the Seventh Five-year Plan in Varanasi.

Measures to Protect Taj Mahal from Pollution

The Government have taken, some special measures to save Taj Mahal from pollution.

A geographical zone around Agra has been notified where new polluting industries and expansion of the present capacities of the existing foundries have been banned. To minimize the effect of pollutants, plan to raise a green belt around Taj has been made by the Department of Environment and is being implemented by the Government of Uttar Pradesh.

The ambient level of sulphur dioxide near Taj is being monitored by National Environmental Engineering Research Institute. Meteorological data are being collected by the Indian Meteorological Department. Indian oil Corporation has set up four ambient monitoring stations for measuring ambient levels of sulphur dioxide at Keethum, Sikandara, Fara and Bharatpur.

The Archaeological Survey has installed very powerful sophisticated sulphur dioxide monitors at Taj and Sikandara. Their sensibility range is 0.5 to 1.0 part per million. The measurement of sulphation rate, analysis of rain water, analysis of particulate matter, cationic analysis, measurement of wind direction and wind velocity, measurement of primary and secondary steps are also being done. In addition, new preservatives are being tried and cleaning of marble by Attanbulgite technique has been introduced. Vigorous measures on way to enssre its conservation, stone by stone is proposed. The tomb underground may soon be closed to public on account of excessive humidity.

Laws to Protect the Environment

In every State a Pollution Control Board has been formed. On November 1, 1980 a Central Pollution Control Board has been created at the national level under the provisions of the Water

(Prevention and Control) Act. This apex body not only advises the Central Government but also Coordinates the activites of all the States Boards.

Pollution Control is a burning problem now-a-days. Many measures have been taken by the Government of India for protecting the environment from pollution. For regulating pollution following legislative measures have been taken by the Government.

India entered in the realm of environmental protection in the previous decade by introducing Water (Prevention and Control of Pollution) Act, 1974. The Act, provides for the prohibition or restriction of the letting out of industrial effluents in streams, wells and other lands. Infringement is punishable with a minimum imprisonment of six months which can be extended up to six years with a fine. However, under the Water Act there is no strict liability for polluting, it is considered an offence only if the pollution is 'knowingly caused or permitted'. Injunctive relief is not provided for, and no private right of action for implementing the right of the State to prosecute violators exists.

Under section 7(1) of the Territorial Waters, continental shelf Exclusive Economic Zone and other Marine Zones Act, 1976 the limit of India's exclusive economic zone is that extending upto 200 nautical miles beyond the territorial waters of the Country, broadly upto 12 nautical miles from the shores. The Central Government has exclusive jurisdiction to preserve and protect the marine environment. The important pollutants of the sea are petroleum and its derivatives, radioactive substances, thermal wastes and suspended solids. The infringement of provisions of notification issued is punishable with imprisonment upto three years or with fine or both.

The Air (Prevention and Control of Pollution) Act, 1981, enacted as a sequence of the decisions taken at the UN Conference on Human Environment, 1972 provides for the prevention, control and abatement of air pollution.

The Air Act regulates all industries in the designated air pollution—control areas as well as industries listed hazardous, irrespective of the location. The statute requires consent only for scheduled industries which operate in the pollution control area. This narrows down the scope of the Act, though some boards have declared the entire state a pollution-control area. One provision empowered State Boards to enter and inspect premises only at reasonable times.

Catastrophic events like the Bhopal gas tragedy and the nuclear accident at chernobyl have helped to focus attention on environmental issues as never before. They have naturally led to the quarries about the existing environmental laws and their adequacy in safeguarding and protecting environment.

BHOPAL GAS TRAGEDY (1984)

The MIC gas leak in Bhopal in 1984 is probably the worst industrial accident which is related to air pollution. Around 2,00,000 Bhopal residents were affected by the leak of poisonous MIC gas from the Union Carbide Pesticide plant there. Atleast 2,500 people were killed and doctors estimated that some 50,000 people have been seriously affected and many go blind.

MIC *i.e.* Methyl-iso-cyanate is a toxic gas used in the manufacture of pesticides. It reacts quickly with water and causes the lungs to swell and eyes to develop cataract. Many died in Bhopal because their lungs had filled with fluid.

Bhopal's victims continue to die. Out of every 3 children born to women who were pregnant on the night of the disaster, only one survived. Out of 1,350 new born babies, 16 were physically deformed and 60 premature births. Deformities include children suffering from Cogenital hearts, holes in arms and impaired eye sight. High levels of thiocyanates were detected in water in Bhopal and continued exposure to this may cause adverse functioning of organs like thyroid, which in turn may affect pregnancy.

The vegetation in an area of 3.5 sq. km. around the Union Carbide factory at Bhopal was severally affected. Leaves bore the burnt of the damages. Consumption of fruit from trees in affected localities—especially ber, mango, papaya and tamarind was avoided for that season. Cultivated plants were more damaged than the wild plants. Plants submerged in water were less affected than the plants exposed to the gas.

CHERNOBYL NUCLEAR MISHAP

On April 28th, 1986, a nuclear reactor meltdown and explosion had occurred at the chernobye nuclear power station, 80 miles north of kiev in USSR. This may be considered the worst disaster in 32 years of commercial atomic power which has caused. Untold death, suffering and environmental and health damage. The dangerous radiation floated towards neighbouring countries.

Around 2,000 people near the plant were reported to have been killed by causes ranging from the initial blast to lethal radiation and about ten thousand may have been evacuated from the endangered region. Radioactive gases and particles have spreadover a vast section of the Soviet union and water supplies for more than 6 million inhabitants of the kiev area are threatened with contamination.

The most frightening part of the nuclear accident was the radiation that spread from the reactor, which was then carried by winds to the neighbouring areas. In the first few hours of the disaster deadly form of iodine and cesium were released into the atmosphere alongwith other highly dangerous radioactive emissions. By the end of the week a large invisible cloud of radiation had spread across Eastern Europe and the shores of the Mediterranean sea. The people who are actually in trouble are those right at the site and local residents who risked exposure to extreme doses of radiation that could cause cerebral haemorrhage, nausea, vomiting and death within hours.

The Union Government enacted a comprehensive 'Environment (Protection) Act, 1986 in May 1986, keeping in view the lacunae and procedural difficulties involved in numerous Central and State Anti-pollution Laws.

The new Act is superior in its approach to environmental problems, compared to the earlier ones. The new Act has a far more comprehensive approach. The definition of pollution is not restricted to air and water pollution only, but covers all kinds of pollutants. Section 6(b) of the Act empowers the Central Government to make rules for 'the maximum allowable limits of concentrations of various environmental pollutants (including noise) for different areas.'

The new act also deals with the hazardous industries. A 'Hazardous substance' is defined in the Act as 'any substance or preparation which, by reason of its chemical or physio-chemical properties or handling, is liable to cause harm to human beings, other living creatures, plant micro-organism, property and enviorment. Moreover, it is now mandatory for a person responsible for the discharge of any hazardous substance in excess of the prescribed norms to immediately inform the concerned authorities and render all possible assistance. Earlier, there was no such responsibility enjoined upon him.

Another outstanding feature is that if a public complaint is made to the Board or Government on a particular issue, the Central Government or any authority or officer in this behalf by that government has to take immediate action in the matter. If no action is taken within 60 days, the complaint can go to the court both against the pollutant and the concerned government.

Whoever fails to comply with or contravenes any of this Act's provisions, rules, orders or directions issued, shall be punished with imprisonment extendable upto five years, or with a fine extendable to Rs. 1,00,000 or with both in respect of each and in case that failure or contravention continues even after the first conviction, with an addition daily fine of Rs. 5,000, if it continues beyond one year the offenders shall be punished for a term extendable to seven years.

The Motor Vehicles rules for Bombay city require petroleum powered Vehicles to have their emissions tested at three month's intervals. A 'Pollution Under Control' (PUC) certificate indicating that the Carbon monoxide emission comply with the State Standards are to be displayed by inspected vehicles. The certificates are obtainable for small fee from gas stations which have purchased the officially approved emission detector and are also prepared to make adjustments in the Carburettor so that vehicles are brought into compliance. The Bombay Motor Vehicles (Amendment) Act, 1984 empowers authority to suspend, after a grace period of 14 days, the registration of vehciles emitting pollutants beyond a certain specified limit.

Minimal National Standards

Any country introducing pollution control legistation at any time would have a stock of industrial units with antiquated or no pollution control and also with old pollution control processes. Pollution control for these units is more difficult as compared to new industries. To expedite establishment of pollution control in the existing group of industry, the Central Pollution Control Board evolved an industry—specific pollution control strategy based on the comprehensive industry document. This document covers numbers, sizes, geographical distribution, material inputs, processes adopted, product mix, by-products recovery, water consumption, various waste streams and their characterization. It particularly deals in prevention and control methods of waste water as against their cost implications. Any specific type of industry must know the extent upto which its effluence must be treated so that it can discharge the same either for irrigation or in rivers, lakes, esturaries and sea depending on its location.

The document evolves the industry-specific Minimal National Standards (MINAS) by evaluating cost of various levels of treatment. The capital cost on this measure is converted in annuity taking relevant interest on capital and depreciation. This annuity is converted into annual burden by adding operational, maintenance, and repair cost for each level of treatment. The level of treatment for which the annual burden remains within three per cent of the turnover is accepted as the appropriate level of treatment, to be installed by each existing specific industry. The concomitant effluent quality is termed MINAS which cannot be relaxed by State pollution control Boards. Stricter effluent quality is prescribed of specific location so demands. The consultant engaged for envolving each of the comprehensive Industry Document is continously guided by an Industry committee.

Any new industry draws attention of the provisions of the water Act only at the stage of discharging the industrial waste water. Hence, at the instance of the Central Pollution Control Board, the Central Government issued an executive order. According to this executive order the central Industrial Licencing Committee allows coversion of letter of intent into industrial licence if the entrepreneur obtains a No Objection Certificate (NOC) from the concerned State Pollution Control Board. In the States where NOC is practised, entrepreneurs of new industries submit six monthly progress reports on commissioning of the industrial plant to the Directorate General of Technical Development (DGTD) of the Central Government alongwith the report on pollution control equipment erection.

Environmental pollution is a matter of global concern. Many countries of the world, especially developed countries have made several enactments for the prevention and control of pollution.

The Federal Republic of Germany is to become one of the first countries to give constitutional status to protection of the environment and nature conservation. It has been decided to include into the objective of the federal constitution, thus, underlying the State's responsibility for maintaining a healthy environment, the effective protection of nature's prudent use of the world's natural resources and repair, of existing damage.

As part of the plans to expand protection of the environment the Government will initiate a standardised code of environmental legislation, introduce compulsory liability, insurance against harm to the environment for all industrial enterprises and promote the manufacture of products qualifying for the 'Blue Angel, mark of environmental comparability.

In taking action of this kind, the Federal Government has the support of the population and industry, even if it places obligations on them. Some 3,000 environment officers are employed in industry to police the environmental regulations. Environmental protection has spawned a whole new growth industry in Federal Republic of Germany.

Constitutional Provisions

The Constitution of India holds the Government responsible under the Directive Principles of State Policy for the protection of environment whereas it also holds the citizens responsible under the Fundamental duties of citizens. Our constitution under Article 48-A provides that 'the State shall endeavour to protect and improve the environment and to safeguard the forests and wild life of the country.' It is one of the fundamental duties of citizen, under Article 51(A) (g), to protect and improve the natural environment including forests, lakes, rivers and wild life and have compassion for living creature.'

The Seventh Plan document recognises the task of bringing on third geographical area of the country under tree-cover as an imperative and prescribes that all possible efforts be made to achieve the target by the turn of the century. This implies that the target of 33 per cent tree-cover which was fixed in 1952 has not yet been attained. The action programme has been detailed in the Prime Minister's broadcast to the nation in January 1985, "continuing deforestation has brought us face to face with a major ecological and socio-economic crisis. The trend must be halted. I propose immediataly to set up a National Wastelands Development Board with the object of bringing five million hectares of land every year under fuelwood and fodder plantation. We shall develop a people's movement for afforestation.

It is heartening that the aims and objective in the forestry sector, which were so long vague and unspecified, have been clealy spelt out. The National Wasteland Development Board (NWDB), that came into being in 1985, works under the overall guidance of the National Land Use and Wasteland Development Council headed by the Prime Minister.

Though the classified as forests in the country is 75 million hectares being 23 per cent of the geopraphical area, a recent report of the National Remote Sensing Agency (NRSA) indicates that only 36 million hectares have good forests. The most significant revelation is the loss of good forests cover in 1975-82 at 15 million hectares per annum.

As a result of deforestation in India firewood has become more costly. Fishermen in Madras find it difficult to make new catamarans (a raft of logs fied together) because wood is scarce and expensive. Bhabhar grass is not available to rope makes in Uttar Pradesh. Acute fodder shortage has affected the cattle feeding in India. Many tribes and millions of other forest dwelling people have been displaced and deprived of their possessions.

Deforestation is posing a serious ecological threat of floods, water shortage, famine, land slides, soil erosion desertification and threat to wild life.

It is estimated that six per cent of land area of the globe is occupied by the jungles, while half the plant and animal species of the world exist in tropical rain forests. Rough estimates of how much of the jungle has already fallen prey to man's whimsical destruction put the figure as 11 million hectares a year. The total requirement of firewood by 2000 A.D. would be around 230 million tonnes. The country now needs 590 million tonnes green fodder and 485 tonnes of dry fodder as against the availability of 350 million tonnes of green fodder and 441 million tonnes of dry fodder. The requirement by 2000 A.D. is estimated to touch 780 million tonnes of green fodder and 600 million tonnes of dry fodder. Thus, the gap between the deforested area and the afforested area would widen unless adequate steps would be taken in time.

For a proper balance in nature, land surface should be forests. Planting trees is known as afforestation.

Within the country and in government agencies there is a new consciousness about the importance of forests, and the rate of afforestation has gone up in recent years.

The Chipko Andolan—the movement to hug trees—was born one morning in March 1973 in the hill town of Gopeswar in Chamoli District of Uttar Pradesh, when representatives of a sports goods factory came to cut 10 trees near Mandal Village. The villagers told them not to do so, but when the contractors persisted, they hugged the trees and the representatives had to return empty-handed.

A few weeks later when the same contractor went to another village, the villagers of Gopeshwar marched to the spot and hugged the trees, which were to be cut. The chipko movement has greatly helped in organing afforestation programmes like the sponsored camps which brought together local villages, students and social workes who have planted over a million trees.

A meeting organised recently by the NWDB has finally evolved a definition for wastelands. Wasteland is now defined as the land which is degraded and presently lying unutilised due to different constraints. The major factors giving rise to wastelands are indiscriminate deforestration, removal of top soil by wind and water, wrong irrigation practices, growth of soil salinity and in some areas shifting cultivation.

Wastelands are further divided into culturable and unculturable wastelands. The former class comprise gullied land, undulating upland, waterlogged land and marshs salt-affected land, shifting cultivation area, sandy area, mining and industrial wasteland pastures and degraded forests. Barren rocky areas, steep slopes, snow-covered and glacia areas constitute the latter. Needless to mention that afforestation programmes are concerned with culturable wastelands only. In the absence of an agreed definition for wastelands their estimates worked out by different agencies vary from 155 to 175 million hectares of degraded forests.

Public opinion has to be moulded so that tree planting is taken as seriously as agriculture and attempts at deforestation are resisted. Recent estimates reveal that the area affected by floods has shot upto 40 million hectares 20 million hectares in 1971.

The country has been estimated to be losing more than six billion tons of top soil every year from the Ganga basin alone.

The public needs to be educated about the positive aspects of forestry also. Grass and fodder trees grown on wastelands will meet the needs of domestic animals. Trees grown on denuded land will meet the needs of household energy. Thus not only afforestation but social forestry also need to be emphasized.

Social forestry is the tree raising programme aimed at meeting the fuel, fodder, fruit and timber needs of the rural population. It is raised by the village people on wastelands. Selected grass with high nutritive value can be grown as fodder for cattle. Medicinal plants and fodder can be raised and community lands can be used for this purpose.

Rural Development Programmes and NSS in schools and colleges have popularised Social Forestry. In addition to these several State Governments of Tamil Nadu, Karnataka, Gujarat, Haryana, and West Bengal have started social forestry projects with financial aid from agencies like US Agency for International Development.

Larger afforestation programme will also provide more rural employment. Better tree-cover will improve the environment, soil and water regime. In short, afforestation should be presented as an important tool of rural development.

The Government has urged the urban development bodies to plant saplings on either side of the road to maintain the ecological balance and provide traffic is-lands with green pastures and beautiful flowers. Growing a green belt, around cities should be a prority, because this will help in minimising air pollution and more oxygen will be added to the atmosphere which is essential for life.

The national programme on Doordarshan has recently featured a couple of good serials on forestry. The regional centres of Doordarshen and AIR should take more initiative in presenting forestry features. Voluntary agencies can make a significant contribution to the programme by convincing the villagers of the need for greening, assisting them in obtaining tree-pattas, bank finance and extension work relating to planting, nurseries, etc. Greenery absorbs the scorching heat of summer and provides cool breeze, adds oxygen to the air, prevents atmospheric pollution, attracts rains and provides shelter to many birds, etc.

The latest official report indicates that the achievement in afforestation was 3.2 million hectares in the first two years of the Seventh Plan against the professed goal of five million hectares every year. The basic object of afforestation is preservation of environment. The penal provisions of the Indian Forest Act formulated as far back as 1927 are out dated. Some States have enacted their own laws and some others have amended the Central Act to suit their needs. In 1976 the National Commission on Agriculture suggested a uniform central legislation for better protection of forests.

Afforestation

Before discussing afforestation we must know what is a forest ?

A forest may be defined as a large uncultivated tract of land covered with trees of different species growing close together. It is also the home for many animals of different species.

A forest may also be defined as an Eco-system. Eco-system in the sturctural and functional unit of nature and it consists of living and non-living communities. Forest is defined as an eco-system because, it is also a unit of nature consisting of the living and non-living communities.

As a measure to protect environment we should also protect forests. Afforestation is emphasized by the Government. Conservation of forests is most essential. Government is taking keen interest in social forestry to preserve our forest wealth.

From time immemorial forests have been immensely assisting human progress and development Primitive agriculture treated forests unfriendly but as men began to domesticate the animal it served as pasture and thus started civilizations.

Forest is an eco-system consisting plants and animals and their environment, Forests are one of the renewable resources. Due to deforestation there are many problems such as changing climate resulting in floods and famine, soil erosion, siltation, destruction of green coverage leading to the increase in surface temperature and oxygen—carbondioxide imbalances, fuel and wood shortage, dust blow and desertification. After the first war of Independence in 1857 our forests were subjected to maximum exploitation. Realising the diminishing forests, the Indian Government in its forest policy announced in 1952 envisaged one third (108 m.h.a.) forest cover of the total geographical area of 329 m.ha. The Fourth Five-Year Plan (1969-74) emphasised the need for increasing the productivity of forests, forests based industries and to develop forests as a support to rural development. The Fifth Five-Year Plan introduced commercial forestry. During the Sixth Five-Year Plan, Central Government introduced Forest (Conservation) Act, 1980 to regulate the use of forest land. The late Prime Minister Smt. Indira Gandhi, however, identified the dangerous process of conversion of fertile lands to wastelands and mooted the idea of wasteland development by afforestation. The Prime Minister Rajiv Gandhi went ahead and he announced the formation of National Waste-Land Development Board (NWDB) and the same was constituted in the year 1985. The NWDB has been reconstituted as National Land Use and Waste Land Development Council chaired by the Prime Minister himself. The main aim of the above bodies is planting of trees in an area of 5 m. ha. every year.

The establishment of National Land Use and Conservation Board and National Land Development Council is a most welcome step at the time of indiscriminate deforestation. Greater emphasis for afforesfation should be on those villages which have already been devastated by ecological imbalances. Agroforesry and social forestry, not only aimed at commercial forestry, be taken up at rapid pace simultaneously educating them about the consequences of deforestation. The Government should also bring laws for the regulation of grazing. Planting specific trees according to the needs of industrial emission around the industrial infrastructural areas and creating a green belt should be strictly implemented to control the pollution.

In its 'State of the World 1988' report, the fifth in a series of annual assessments, the World Watch Institute lauded the Government of India for giving forestry prominence in the Seventh Plan Programmes.

During the Sixth Five-Year Plan special stress was laid on monitoring and control of Environmental pollution. A programme 'Pollution Control at Source' was introduced. In the seventh plan protection of environment received special attention of the government.

Emphasis on Environment in Seventh Plan

According to the Seventh Plan Document, the problems encountered in the field of environment in India are due to conditions of poverty and under development as also the negative effects of development programmes which have been badly planned or badly implemented. The whole, planning process is aimed at development and the removal of poverty. The need to improve the conditions of our people is pressing, under this pressure many concerned with development activities lose sight of environmental and ecological imperatives.

Realisation concerning these aspects has been with us for only a relatively short period of time, about a decade and a half. The damage being done to the environment, because of the large size of the population and its increase, and scale of developmental activities, is of such magnitude that urgent remedial measures are called for. Official and Voluntary agencies must work together to create the needed awarences; indeed, environment is all pervasive, and the success of our efforts in this area will ultimately call for the involvement of the entire population at all levels.

A total of Rs. 427.91 crore has been allocated for environmental protection during the Seventh Plan period (1985-90). Of this, Rs. 350 crores is in the central sector, Rs. 75.71 crore in the State sector and Rs. 2.20 crore in the Union Territory Sector. Out of Rs. 350 Crores in the Central Sector, Rs. 240 crore has been earmarked for the Ganga Action Plan.

The Seventh Plan strategy on environmental goals relates to :

(*a*) Institutionalising the process of integrating environmental management and development.

(*b*) Inducing organisations at the Central, State and local levels to incorporate environmental safegards in their plans and programmes.

(*c*) Establishing a strong science and technology base for environmental research and development, demonstration, and extension activities.

(*d*) Strengthening mechanism for ensuring mechanism for corrective action with regard to environmental degradation that has already taken place.

(*e*) Creative awareness among the people for effective environmental management by utilising existing institutional framework and seeking full co-operation of more than 200 non-governmental organisations of voluntary agéncies.

Pollution Control Measures Taken by Government Organisations

Different Government organisations such as coal India, Railways, Oil and Natural Commission are taking keen interest in Pollution Control.

Coal India is aware of the adverse impact of the mining operations on the environment. Master Plan for environmental management for each coalfield is being drawn. Environmental Management Plan (EMP) is now an integral part of each project report.

Steps being taken by Indian Railways to improve the environment by switching over to pollution free traction. The Oil and Natural Gas Commission (ONGC) is making arrangements for treating the waste water by mobile effluent treatment plant. Water at the wells is contaminated with HDS, burnt oil from internal combustion engines drilling mud chemicals such as cutch, Ferrochrome-Lingo-Sulphonate, Chrome-Lingo Sulphonate, Caustic Soda, Calcium Chloride, Sodium Chloride, Potassium Dichromate,

Baryte, etc. in different proportions. A typical composition of the waste water at the sites is as under :

Oil—1000 to 2000 PPM

Clay (coloidal) 1000—2000 PPM

After treatment, properties of treated water should be as follows :

(*a*) Oil less then — 10 PPM

(*b*) BOD Maximum — 30 PPM

(*c*) COD maximum — 250 PPM

(*d*) Suspended solid — 100 PPM

(*e*) pH — 5.5 to 9.0

The treated effluent water is to be monitored continously.

International Co-operation for Pollution Control

The environment today is endangered from industrial pollutants, indiscriminate deforestation, vehicular emissions, noise, etc. Environmental pollution has attracted global attention. Every country is taking interest to combat pollution through legislative as well other control measures. Different international agencies are working for the prevention and control of pollution. There should be Global environment movement because environment all over world is endangered by pollution. The world meteorological organisation (WMO) by co-ordination of its meteorological studies with health studies conducted by WHO can evolve some standards for stratosphere, International Civil Aviations Organisation (ICAO) may mail regulation to deal this. Similarly the control of the use of oceans for the deliberate waste disposal could be best negotiated by framing regulations by Intergovernmental Maritime Consultative Organisation in consultation with WHO and FAO who will guide it in relation to health and fish aspects respectively.

Though these are recommendatory bodies but can, evaluate environment periodically and influence public opinion in favour of action.

The lack of tradition and capabilities of enforcement by International Organisations argues foy strong national legislation. There is much to be gained by the exchanging informations and experiences with respect to environmental control measures. India has been benefited by the international co-operation in the field of pollution control. Agencies like UN Environment Programme (UNEP), International Union for the Conservation of Nature (IUCN), South Asia co-operative Environment Programme (SACEP), UNIDO, UNESCO, etc. have extended co-operation to India in environmental protection. The Pollution Control Research Institute (PCRI), Hardwar, in collaboration with the United Nations and several international scientific organisations and will hold a major international conference on environmental impact assessment in the end of the year. PCRI also held a national workshop at Hardwar on industrial environment which brought together national and international experts to discuss how best the control and manage air, water and soil pollution from different industrial activities. Global Environmental Monitoring System of WHO is working in India in collaboration with the National Environmental Engineering Research Institution on the problem of pollution.

8

ENVIRONMENTAL EDUCATION

Environment education can be defined as a process of learning about the existing situation through which sufficient knowledge can be gained to understand environmental problems and contribute towards solving them.

Environmental education does not stop with acquiring information about environment but also helps to acquire attitudes and values conducive to environment protection and understanding of interdependence of nature and people.

Environmental education provides :

(*a*) A comprehensive knowledge with working of Nature and Environment.

(*b*) An experience in valuing Environment Quality.

(*c*) An understanding of the impact of personal choices of actions on environmental quality.

(*d*) A source of guidance to the people to act as more responsible citizens with an increased civic sense.

In the Moscow Conference of 1987 the need for international co-operation in environmental education was strongly stressed.

The chief objectives of environmental education are that individuals and social groups should acquire awareness and knowledge. Develop attitudes and skills and abilities and participate in solving real life problems with a practical bias for developing a healthy environment around. The perspective should be integrated, interdisciplinary in character.

Environmental education includes awareness, real life situations, conservation and sustainable development. This has to be matched with the needs of the primary, lower secondary, higher secondary and adult education. Awareness includes making the individual conscious about physical, social and aesthetic aspects of the environment. Creative awareness is the one of the main environmental goals stated in the Seventh Plan document. According to the document stress should be laid on developing creative awareness among the people for environmental management.

The National Museum of Natural History has been spearheading the awareness campaign since its inception in 1978 and today boasts of an annual visitor level of about 1.3 lakhs—an impressive figure in the Indian context. The thrust of the institution is aimed at the younger generation and school children are encouraged to visit both in groups and individually. The various kits provided by the museum are geared to cognitive participation and the fact that these educational kits are issue in both English and Hindi extends the spectrum of the audience that can be covered.

Human being is interlinked with the life support system which in itself has five elements : Air, water, land, flora and fauna. These elements have a dynamic, continuing and living relationship, and the consciousness that humankind, being the most dominant of the species, has a very major responsibility.

The main objectives set for the global community at the Stockholm conference were—

(*a*) to increase our knowledge of the environment and

(*b*) to safe guard and enhance the environment for present and future generations, spread of environmental education, creation of awareness and dissemination of information are very crucial to protect environment effectively.

Environmental education must create an awareness of the economic, political and ecological interdependence of the modern world so as to enhance a spirit of responsibility and solidarity among nations and peoples. This is an essential pre-requisite for resolving serious environmental problem at Global level.

The Ministry of Environment has launched a National Environmental Awareness Campaign since July 1986. The objective is to create environmental awareness at national level for varied target groups.

The primary focus of the campaign has been on the students, teachers and general public. Other key target groups have been people's selected representatives, professionals (Engineers, corporators, Managers, Lawyers, etc.) and media personnel, especially Journalists. Several Universities, Voluntary agencies, non-governmental organisations/ professional associations and societies, nature clubs, community action groups, etc. from virtually every State and Union Territory have been involved in organising various environmentally related programmes. These programmes, relate to seminars, workshops, training programmes film, essay competition, rallies, etc. In this context the 100 day, 1500 km. long march undertaken recently by various Indian environment groups through six States 'to save the western ghats' is significant and like the 'chipko Movement' in the north, it has helped to bring an awareness among the people to the dangers inherent mindless exploitation of forests and the disruption of ecological balance. All possible media and target groups are being involved in the campaign. Fortnightly TV programmes on environment are being telecast through doordarshan since August 1986. The emphasis, during 1986 has mainly been on creative awareness on general aspects of environment.

The 'citizens Against Pollution' alongwith several other student groups social service, organisation organised a 'Race Against Pollution' in Hyderabad in September 1986 to highlight the need to preserve our ecological balance and project the environment. It was a successful venture and helped the citizen to realize the gravity of the problem caused by pollution.

To develop an awareness among the masses, Akashvani, has conducted a unique experiment in Karnataka State with corporation of State's Forest department.

'Akashvani', Bangalore has broadcasted thirteen programmes of 'Nisang Sampada' since August 17, 1987. This programme deals with the problems of environment, role of flora and fauna, importance of forests, pollution control, and the conservation of eco-systems *e g.*, lakes, ponds, etc. This programme also imparted knowledge regarding the importance of social forestry and grave consequence of deforestation. Forest department has also prepared cassettes of the 'Nisang Sampada' programme. People from every cross section of society took interest in this broadcast and enrolled themselves under this programme. This programme helped a lot in developing creative awareness among the people.

Students in schools should be educated about the necessity for personal and environmental cleanliness. The NCERT has made significant progress in this field by including ample material in the field of environment right from class one onwards.

Environmental education is a very vast field dealing with diverse aspects like, human settlements, land, forests, mountains, Island, coastal and other eco-systems, soil, water, Genetic resources, wild life, marine resources, etc. Environmental education must motivate the youth of the country to study the range of environment problems facing India today and to project responsible solutions that are consistent with the economic political and social temper of our country and are practicable. We celebrate on 5 June, World Environment Day. The main object of this is to make people aware of the harms caused to environment in the world. Conferences and exhibitions are organised on this occasion, newspapers publish special articles on environment,

TV and Radio bring different programmes on the problems of environment to the people. The sole object behind all these is to develop among people an awareness for the clean environment. We must protect our environment to save the future.

For the protection of our environments besides government machinery other non-governmental organisation and voluntary agencies should extend full co-operation in organising public awareness activities and thus work for the great cause of protection of environment.

INDEX